An Enquiry Concerning the Principles of Natural Knowledge

ALFRED NORTH WHITEHEAD

"PHILONOUS. *I am not for imposing any sense on your words: you are at liberty to explain them as you please. Only, I beseech you, make me understand something by them.*"

BERKELEY,
The First Dialogue between
Hylas and Philonous.

Dover Publications, Inc., New York

TO

ERIC ALFRED WHITEHEAD

ROYAL FLYING CORPS

November 27, 1898 to March 13, 1918

Killed in action over the Forêt de Gobain
giving himself that the city of his vision
may not perish.

The music of his life was without discord,
perfect in its beauty.

This Dover edition, first published in 1982, is
an unabridged and unaltered republication of the
second (1925) edition of the work originally
published in 1919.

Manufactured in the United States of America
Dover Publications, Inc.
180 Varick Street
New York, N.Y. 10014

**Library of Congress Cataloging in Publication
Data**

Whitehead, Alfred North, 1861-1947.
    An enquiry concerning the principles of natu-
ral knowledge.

    Reprint. Originally published: 2nd ed. Cam-
bridge : Cambridge University Press, 1925.
    1. Science-Philosophy. 2. Knowledge, Theory
of. 3. Space and time. I. Title.
Q175.W6   1982   501   82-7259
ISBN 0-486-24343-5 (pbk.)   AACR2

PREFACE

THERE are three main streams of thought which are relevant to the theme of this enquiry; they may, with sufficient accuracy, be termed the scientific, the mathematical, and the philosophical movements.

Modern speculative physics with its revolutionary theories concerning the natures of matter and of electricity has made urgent the question, What are the ultimate data of science? It is in accordance with the nature of things that mankind should find itself acting and should then proceed to discuss the rationale of its activities. Thus the creation of science precedes the analysis of its data and can even be accompanied by the acceptance of faulty analyses, though such errors end by warping scientific imagination.

The contributions of mathematics to natural science consist in the elaboration of the general art of deductive reasoning, the theory of quantitative measurement by the use of number, the theory of serial order, of geometry, of the exact measurement of time, and of rates of change. The critical studies of the nineteenth century and after have thrown light on the nature of mathematics and in particular on the foundations of geometry. We now know many alternative sets of axioms from which geometry can be deduced by the strictest deductive reasoning. But these investigations concern geometry as an abstract science deduced from hypothetical premises. In this enquiry we are concerned with geometry as a physical science. How is space rooted in experience?

The modern theory of relativity has opened the possibility of a new answer to this question. The successive labours of Larmor, Lorentz, Einstein, and Minkovski have opened a new world of thought as to the relations of space and time to the ultimate data of perceptual knowledge. The present work is largely concerned with providing a physical basis for the more modern views which have thus emerged. The whole investigation is based on the principle that the scientific concepts of space and time are the first outcome of the simplest generalisations from experience, and that they are not to be looked for at the tail end of a welter of differential equations. This position does not mean that Einstein's recent theory of general relativity and of gravitation is to be rejected. The divergence is purely a question of interpretation. Our time and space measurements may in practice result in elaborate combinations of the primary methods of measurement which are explained in this work. For example, the theory of gravitational matter may involve the theory of 'vagrant solids' which is pointed out as a subject for investigation in article 39, but not developed. It has certainly resulted from Einstein's investigations that a modification of the gravitational law, of an order of magnitude which is v^2/c^2 of the main effect [v being the velocity of the matter and c that of light], will account for the more striking outstanding difficulties otherwise unexplained by the law of gravitation. This is a remarkable discovery for which the utmost credit is due to the author. Now that the fact is known, it is easy to see that it is the sort of modification which on the simple electromagnetic theory of relativity is likely to be required for this law. I have however been anxious to disentangle the consideration of the main positions of this enquiry from theories designed to explain special laws of nature.

Also at the date of writing the evidence for some of the
consequences of Einstein's theory is ambiguous and
even adverse. In connection with the theory of relativity I have received suggestive stimulus from Dr L.
Silberstein's *Theory of Relativity*, and from an important Memoir* by Profs. E. B. Wilson and G. N. Lewis.

The discussion of the deduction of scientific concepts
from the simplest elements of our perceptual knowledge
at once brings us to philosophical theory. Berkeley,
Hume, Kant, Mill, Huxley, Bertrand Russell and
Bergson, among others, have initiated and sustained
relevant discussions. But this enquiry is touched by
only one side of the philosophical debate. We are concerned only with Nature, that is, with the object of
perceptual knowledge, and not with the synthesis of
the knower with the known. This distinction is exactly
that which separates natural philosophy from meta-
physics. Accordingly none of our perplexities as to
Nature will be solved by having recourse to the consideration that there is a mind knowing it. Our theme
is the coherence of the known, and the perplexity which
we are unravelling is as to what it is that is known.
In matters philosophic the obligations of an author to
others usually arise from schools of debate rather than
from schools of agreement. Also such schools are the
more important in proportion as assertion and retort do
not have to wait for the infrequent opportunities of
formal publication, hampered by the formidable permanence of the printed word. At the present moment
England is fortunate in this respect. London, Oxford
and Cambridge are within easy reach of each other, and

---

* 'The Space-Time Manifold of Relativity.' *Proc. of the
Amer. Acad. of Arts and Sciences*, vol. XLVIII, 1912.

provide a common school of debate which rivals schools of the ancient and medieval worlds. Accordingly I have heavy obligations to acknowledge to Bertrand Russell, Wildon Carr, F. C. Schiller, T. P. Nunn, Dawes Hicks, McTaggart, James Ward, and many others who, amid their divergencies of opinion, are united in the candid zeal of their quest for truth.

It is quite unnecessary to draw attention to the incompleteness of this investigation. The book is merely an enquiry. It raises more difficulties than those which it professes to settle. This is inevitable in any philosophical work, however complete. All that one can hope to do is to settle the right sort of difficulties and to raise the right sort of ulterior questions, and thus to accomplish one short step further into the unfathomable mystery.

Memories are short, and perhaps it is not inapt to put on record circumstances common to the life of all England during years of war. The book is the product of intervals of leisure amid pressing occupation, a refuge from immediate fact. It has been thought out and written amid the sound of guns—guns of Kitchener's army training on Salisbury Plain, guns on the Somme faintly echoing across the Sussex coast: some few parts composed to pass times of expectation during air-raids over London, punctuated by the sound of bombs and the answer of artillery, with argument clipped by the whirr of aeroplanes. And through the land anxiety, and at last the anguish which is the price of victory.

A. N. W.

April 20, 1919

PREFACE TO SECOND EDITION

SINCE the publication of the first edition of this book in 1919, the various topics contained in it have been also considered by me in *The Concept of Nature* (Camb. Univ. Press, 1920) and in *The Principle of Relativity* (Camb. Univ. Press, 1922). I hope in the immediate future to embody the standpoint of these volumes in a more complete metaphysical study.

A few notes have been appended to this edition to elucidate obscurities, and a few slips in the text have been corrected.

A. N. W.

TRINITY COLLEGE,
CAMBRIDGE
August, 1924

CONTENTS

PART I

THE TRADITIONS OF SCIENCE

CHAPTER I. MEANING

CHAPTER II. THE FOUNDATIONS OF DYNAMICAL PHYSICS

CHAPTER III. SCIENTIFIC RELATIVITY

CHAPTER IV. CONGRUENCE

CONTENTS

PART II

THE DATA OF SCIENCE

CHAPTER V. THE NATURAL ELEMENTS

CHAPTER VI. EVENTS

CHAPTER VII. OBJECTS

PART III

THE METHOD OF EXTENSIVE ABSTRACTION

CHAPTER VIII. PRINCIPLES OF THE METHOD OF EXTENSIVE ABSTRACTION

PART IV

THE THEORY OF OBJECTS

CHAPTER XIV. THE LOCATION OF OBJECTS

CHAPTER XV. MATERIAL OBJECTS

CHAPTER XVI. CAUSAL COMPONENTS

CHAPTER XVII. FIGURES

CHAPTER XVIII. RHYTHMS

PART I

THE TRADITIONS OF SCIENCE

CHAPTER I

MEANING

1. Traditional Scientific Concepts. 1·1 What is a physical explanation? The answer to this question, even when merely implicit in the scientific imagination, must profoundly affect the development of every science, and in an especial degree that of speculative physics. During the modern period the orthodox answer has invariably been couched in terms of Time (flowing equably in measurable lapses) and of Space (timeless, void of activity, euclidean), and of Material in space (such as matter, ether, or electricity).

The governing principle underlying this scheme is that extension, namely extension in time or extension in space, expresses disconnection. This principle issues in the assumptions that causal action between entities separated in time or in space is impossible and that extension in space and unity of being are inconsistent. Thus the extended material (on this view) is essentially a multiplicity of entities which, as extended, are diverse and disconnected. This governing principle has to be limited in respect to extension in time. The same material exists at different times. This concession introduces the many perplexities centering round the notion of change which is derived from the comparison of various states of self-identical material at different times.

1·2 The ultimate fact embracing all nature is (in this traditional point of view) a distribution of material throughout all space at a durationless instant of time, and another such ultimate fact will be another distribution of the same material throughout the same space at another durationless instant of time. The difficulties of this extreme statement are evident and were pointed out even in classical times when the concept first took shape. Some modification is evidently necessary. No room has been left for velocity, acceleration, momentum, and kinetic energy, which certainly are essential physical quantities.

We must therefore in the ultimate fact, beyond which science ceases to analyse, include the notion of a state of change. But a state of change at a durationless instant is a very difficult conception. It is impossible to define velocity without some reference to the past and the future. Thus change is essentially the importation of the past and of the future into the immediate fact embodied in the durationless present instant.

This conclusion is destructive of the fundamental assumption that the ultimate facts for science are to be found at durationless instants of time.

1·3 The reciprocal causal action between materials *A* and *B* is the fact that their states of change are partly dependent on their relative locations and natures. The disconnection involved in spatial separation leads to reduction of such causal action to the transmission of stress across the bounding surface of contiguous materials. But what is contact? No two points are in contact. Thus the stress across a surface necessarily acts on some bulk of the material enclosed inside. To say that the stress acts on the immediately contiguous material is to assert infinitely small volumes. But there are no such things, only smaller and smaller volumes. Yet (with this point of view) it cannot be meant that the surface acts on the interior.

Certainly stress has the same claim to be regarded as an essential physical quantity as have momentum and kinetic energy. But no intelligible account of its meaning is to be extracted from the concept of the continuous distribution of diverse (because extended) entities through space as an ultimate scientific fact. At some stage in our account of stress we are driven to the concept of any extended quantity of material as a single unity whose nature is partly explicable in terms of its surface stress.

1·4 In biology the concept of an organism cannot be expressed in terms of a material distribution at an instant. The essence of an organism is that it is one thing which functions and is spread through space. Now functioning takes time. Thus a biological organism is a unity with a spatio-temporal extension which is of the essence of its being. This biological conception is obviously incompatible with the traditional ideas. This argument does not in any way depend on the assumption that biological phenomena belong to a different category to other physical phenomena. The essential point of the criticism on traditional concepts which has occupied us so far is that the concept of unities, functioning and with spatio-temporal extensions, cannot be extruded from physical concepts. The only reason for the introduction of biology is that in these sciences the same necessity becomes more clear.

1·5 The fundamental assumption to be elaborated in the course of this enquiry is that the ultimate facts of nature, in terms of which all physical and biological explanation must be expressed, are events connected by their spatio-temporal relations, and that these relations are in the main reducible to the property of events that they can contain (or extend over) other events which are parts of them. In other words, in the place of emphasising space and time in their capacity of disconnecting, we shall build up an account of their complex essences as derivative from the ultimate ways in which those things, ultimate in science, are interconnected. In this way the data of science, those concepts in terms of which all scientific explanation must be expressed, will be more clearly apprehended. But before proceeding to our constructive task, some further realisation of the perplexities introduced by the traditional concepts is necessary.

2. *Philosophic Relativity.* 2·1 The philosophical principle of the relativity of space means that the properties of space are merely a way of expressing relations between things ordinarily said to be 'in space.' Namely, when two things are said to be 'both in space' what is meant is that they are mutually related in a certain definite way which is termed 'spatial.' It is an immediate consequence of this theory that all spatial entities such as points, straight lines and planes are merely complexes of relations between things or of possible relations between things.

For consider the meaning of saying that a particle P is at a point Q. This statement conveys substantial information and must therefore convey something more than the barren assertion of self-identity 'P is P.' Thus what must be meant is that P has certain relations to other particles P', P'', etc., and that the abstract possibility of this group of relations is what is meant by the point Q.

The extremely valuable work on the foundations of geometry produced during the nineteenth century has proceeded from the assumption of points as ultimate given entities. This assumption, for the logical purpose of mathematicians, is entirely justified. Namely the mathematicians ask, What is the logical description of relations between points from which all geometrical theorems respecting such relations can be deduced? The answer to this question is now practically complete; and if the old theory of absolute space be true, there is nothing more to be said. For points are ultimate simple existents, with mutual relations disclosed by our perceptions of nature.

But if we adopt the principle of relativity, these investigations do not solve the question of the foundations of geometry. An investigation into the foundations of geometry has to explain space as a complex of relations between things. It has to describe what a point is, and has to show how the geometric relations between points issue from the ultimate relations between the ultimate things which are the immediate objects of knowledge. Thus the starting point of a discussion on the foundations of geometry is a discussion of the character of the immediate data of perception. It is not now open to mathematicians to assume *sub silentio* that points are among these data.

2·2 The traditional concepts were evidently formed round the concept of absolute space, namely the concept of the persistent ultimate material distributed among the persistent ultimate points in successive configurations at successive ultimate instants of time. Here 'ultimate' means 'not analysable into a complex of simpler entities.' The introduction of the principle of relativity adds to the complexity—or rather, to the perplexity—of this conception of nature. The statement of general character of ultimate fact must now be amended into 'persistent ultimate material with successive mutual ultimate relations at successive ultimate instants of time.'

Space issues from these mutual relations of matter at an instant. The first criticism to be made on such an assertion is that it is shown to be a metaphysical fairy tale by any comparison with our actual perceptual knowledge of nature. Our knowledge of space is based on observations which take time and have to be successive, but the relations which constitute space are instantaneous. The theory demands that there should be an instantaneous space corresponding to each instant, and provides for no correlation between these spaces; while nature has provided us with no apparatus for observing them.

2·3 It is an obvious suggestion that we should amend our statement of ultimate fact, as modified by the acceptance of relativity. The spatial relations must now stretch across time. Thus if P, P', P'', etc. be material particles, there are definite spatial relations connecting P, P', P'', etc. at time t_1 with P, P', P'', etc. at time t_2, as well as such relations between P and P' and P'', etc. at time t_1 and such relations between P and P' and P'', etc. at time t_2. This should mean that P at time t_2 has a definite position in the spatial configuration constituted by the relations between P, P', P'', etc. at time t_1.

For example, the sun at a certain instant on Jan. 1st, 1900 had a definite position in the instantaneous space constituted by the mutual relations between the sun and the other stars at a definite instant on Jan. 1st, 1800. Such a statement is only understandable (assuming the traditional concept) by recurring to absolute space and thus abandoning relativity; for otherwise it denies the completeness of the instantaneous fact which is the essence of the concept. Another way out of the difficulty is to deny that space is constituted by the relations of P, P', P'', etc., at an instant, and to assert that it results from their relations throughout a duration of time, which as thus prolonged in time are observable.

As a matter of fact it is obvious that our knowledge of space does result from such observations. But we are asking the theory to provide us with actual relations to be observed. This last emendation is either only a muddled way of admitting that 'nature at an instant' is not the ultimate scientific fact, or else it is a yet more muddled plea that, although there is no possibility of correlations between distinct instantaneous spaces, yet within durations which are short enough such non-existent correlations enter into experience.

2·4 The persistence of the material lacks any observational guarantee when the relativity of space is admitted into the traditional concept. For at one instant there is instantaneous material in its instantaneous space as constituted by its instantaneous relations, and at another instant there is instantaneous material in its instantaneous space. How do we know that the two cargoes of material which load the two instants are identical? The answer is that we do not perceive isolated instantaneous facts, but a continuity of existence, and that it is this observed continuity of existence which guarantees the persistence of material. Exactly so; but this gives away the whole traditional concept. For a 'continuity of existence' must mean an unbroken duration of existence. Accordingly it is admitted that the ultimate fact for observational knowledge is perception through a duration; namely, that the content of a specious present, and not that of a durationless instant, is an ultimate datum for science.

2·5 It is evident that the conception of the instant of time as an ultimate entity is the source of all our difficulties of explanation. If there are such ultimate entities, instantaneous nature is an ultimate fact.

Our perception of time is as a duration, and these instants have only been introduced by reason of a supposed necessity of thought. In fact absolute time is just as much a metaphysical monstrosity as absolute space. The way out of the perplexities, as to the ultimate data of science in terms of which physical explanation is ultimately to be expressed, is to express the essential scientific concepts of time, space and material as issuing from fundamental relations between events and from recognitions of the characters of events. These relations of events are those immediate deliverances of observation which are referred to when we say that events are spread through time and space.

3. *Perception.* 3·1 The conception of one universal nature embracing the fragmentary perceptions of events by one percipient and the many perceptions by diverse percipients is surrounded with difficulties. In the first place there is what we will call the 'Berkeleyan Dilemma' which crudely and shortly may be stated thus: Perceptions are in the mind and universal nature is out

of the mind, and thus the conception of universal
nature can have no relevance to our perceptual life.
This is not how Berkeley stated his criticism of mater-
ialism; he was thinking of substance and matter. But
this variation is a detail and his criticism is fatal to
any of the traditional types of 'mind-watching-things'
philosophy, even if those things be events and not
substance or material. His criticisms range through
every type of sense-perception, though in particular
he concentrates on Vision.

3·2 "*Euphranor**. Tell me, Alciphron, can you discern the
doors, windows, and battlements of that same castle?
Alciphron. I cannot. At this distance it seems only a small
round tower.
Euph. But I, who have been at it, know that it is no small round
tower, but a large square building with battlements and
turrets, which it seems you do not see.
Alc. What will you infer from thence?
Euph. I would infer that the very object which you strictly and
properly perceive by sight is not that thing which is several
miles distant.
Alc. Why so?
Euph. Because a little round object is one thing, and a great
square object is another. Is it not so?
Alc. I cannot deny it.
Euph. Tell me, is not the visible appearance alone the proper
object of sight?
Alc. It is.
What think you now (said *Euphranor*, pointing towards
the heavens) of the visible appearance of yonder planet?
Is it not a round luminous flat, no bigger than a six-pence?
Alc. What then?
Euph. Tell me then, what you think of the planet itself? Do

* *Alciphron*, The Fourth Dialogue, Section 10.

you not conceive it to be a vast opaque globe, with several unequal risings and valleys?

Alc. I do.

Euph. How can you therefore conclude that the proper object of your sight exists at a distance?

Alc. I confess I do not know.

Euph. For your further conviction, do but consider that crimson cloud. Think you that, if you were in the very place where it is, you would perceive anything like what you now see?

Alc. By no means. I should perceive only a dark mist.

Euph. Is it not plain, therefore, that neither the castle, the planet, nor the cloud, which you see here, are those real ones which you suppose exist at a distance?"

3·3 Now the difficulty to be faced is just this. We may not lightly abandon the castle, the planet, and the crimson cloud, and hope to retain the eye, its retina, and the brain. Such a philosophy is too simple-minded —or at least might be thought so, except for its wide diffusion.

Suppose we make a clean sweep. Science then becomes a formula for calculating mental 'phenomena' or 'impressions.' But where is science? In books? But the castle and the planet took their libraries with them.

No, science is in the minds of men. But men sleep and forget, and at their best in any one moment of insight entertain but scanty thoughts. Science therefore is nothing but a confident expectation that relevant thoughts will occasionally occur. But by the bye, what has happened to time and space? They must have gone after the other things. No, we must distinguish: space has gone, of course; but time remains as relating the succession of phenomena. Yet this won't do; for this succession is only known by recollection, and recollection is subject to the same criticism as that applied by Berkeley to the castle, the planet, and the cloud. So after all, time does evaporate with space, and in their departure 'you' also have accompanied them; and I am left solitary in the character of a void of experience without significance.

3·4 At this point in the argument we may break off, having formed a short catalogue of the sort of considerations which lead from the Berkeleyan dilemma to a complete scepticism which was not in Berkeley's own thought.

There are two types of answer to this sceptical descent. One is Dr Johnson's. He stamped his foot on a paving-stone, and went on his way satisfied with its reality. A scrutiny of modern philosophy will, if I am not mistaken, show that more philosophers should own Dr Johnson as their master than would be willing to acknowledge their indebtedness.

The other type of answer was first given by Kant. We must distinguish between the general way he set about constructing his answer to Hume, and the details of his system which in many respects are highly disputable. The essential point of his method is the assumption that 'significance' is an essential element in concrete experience. The Berkeleyan dilemma starts with tacitly ignoring this aspect of experience, and thus with putting forward, as expressing experience, conceptions of it which have no relevance to fact. In the light of Kant's procedure, Johnson's answer falls into its place; it is the assertion that Berkeley has not correctly expounded what experience in fact is.

Berkeley himself insists that experience is significant, indeed three-quarters of his writings are devoted to enforcing this position. But Kant's position is the converse of Berkeley's, namely that significance is experience. Berkeley first analyses experience, and then expounds his view of its significance, namely that it is God conversing with us. For Berkeley the significance is detachable from the experience. It is here that Hume came in. He accepted Berkeley's assumption that experience is something given, an impression, without essential reference to significance, and exhibited it in its bare insignificance. Berkeley's conversation with God then becomes a fairy tale.

3·5 What is 'significance'? Evidently this is a fundamental question for the philosophy of natural knowledge, which cannot move a step until it has made up its mind as to what is meant by this 'significance' which is experience.

'Significance' is the relatedness of things. To say that significance is experience, is to affirm that perceptual knowledge is nothing else than an apprehension of the relatedness of things, namely of things in their relations and as related. Certainly if we commence with a knowledge of things, and then look around for their relations we shall not find them. 'Causal connection' is merely one typical instance of the universal ruin of relatedness. But then we are quite mistaken in thinking that there is a possible knowledge of things as unrelated. It is thus out of the question to start with a knowledge of things antecedent to a knowledge of their relations. The so-called properties of things can always be expressed as their relatedness to other things unspecified, and natural knowledge is exclusively concerned with relatedness.

3·6 The relatedness which is the subject of natural

knowledge cannot be understood without reference to the general characteristics of perception. Our perception of natural events and natural objects is a perception from within nature, and is not an awareness contemplating all nature impartially from without. When Dr Johnson 'surveyed mankind from China to Peru,' he did it from Pump Court in London at a certain date. Even Pump Court was too wide for his peculiar *locus standi*; he was really merely conscious of the relations of his bodily events to the simultaneous events throughout the rest of the universe. Thus perception involves a percipient object, a percipient event, the complete event which is all nature simultaneous with the percipient event, and the particular events which are perceived as parts of the complete event. This general analysis of perception will be elaborated in Part II. The point here to be emphasised is that natural knowledge is a knowledge from within nature, a knowledge 'here within nature' and 'now within nature,' and is an awareness of the natural relations of one element in nature (namely, the percipient event) to the rest of nature. Also what is known is not barely the things but the relations of things, and not the relations in the abstract but specifically those things as related.

Thus Alciphron's vision of the planet is his perception of his relatedness (i.e. the relatedness of his percipient event) to some other elements of nature which as thus related he calls the planet. He admits in the dialogue that certain other specified relations of those elements are possible for other percipient events. In this he may be right or wrong. What he directly knows is his relation to some other elements of the universe—namely, I, Alciphron, am located in my percipient event 'here and now' and the immediately perceived appearance of the planet is for me a characteristic of another event 'there and now.' In fact perceptual knowledge is always a knowledge of the relationship of the percipient event to something else in nature. This doctrine is in entire agreement with Dr Johnson's stamp of the foot by which he realised the otherness of the paving-stone.

3·7 The conception of knowledge as passive contemplation is too inadequate to meet the facts. Nature is ever originating its own development, and the sense of action is the direct knowledge of the percipient event as having its very being in the formation of its natural relations. Knowledge issues from this reciprocal insistence between this event and the rest of nature, namely relations are perceived in the making and because of the making. For this reason perception is always at the utmost point of creation. We cannot put ourselves back to the Crusades and know their events while they were happening. We essentially perceive our relations with nature because they are in the making. The sense of action is that essential factor in natural knowledge which exhibits it as a self-knowledge enjoyed by an element of nature respecting its active relations with the whole of nature in its various aspects. Natural knowledge is merely the other side of action. The forward moving time exhibits this characteristic of experience, that it is essentially action. This passage of nature—or, in other words, its creative advance—is its fundamental characteristic; the traditional concept is an attempt to catch nature without its passage.

3·8 Thus science leads to an entirely incoherent philosophy of perception in so far as it restricts itself to the ultimate datum of material in time and space, the spatio-temporal configuration of such material being the object of perception. This conclusion is no news to philosophy, but it has not led to any explicit re-organisation of the concepts actually employed in science. Implicitly, scientific theory is shot through and through with notions which are frankly inconsistent with its explicit fundamental data.

This confusion cannot be avoided by any kind of theory in which nature is conceived simply as a complex of one kind of inter-related elements such as either persistent things, or events, or sense-data. A more elaborate view is required of which an explanation will be attempted in the sequel. It will suffice here to say that it issues in the assertion that all nature can (in many diverse ways) be analysed as a complex of things; thus all nature can be analysed as a complex of events, and all nature can be analysed as a complex of sense-data. The elements which result from such analyses, events, and sense-data, are aspects of nature of funda-mentally different types, and the confusions of scientific theory have arisen from the absence of any clear re-cognition of the distinction between relations proper to one type of element and relations proper to another type of element. It is of course a commonplace that elements of these types are fundamentally different. What is here to be insisted on is the way in which this commonplace truth is important in yielding an analysis of the ultimate data for science more elaborate than that of its current tradition. We have to remember that while nature is complex with time-less subtlety, human thought issues from the simple-mindedness of beings whose active life is less than half a century.

CHAPTER II

THE FOUNDATIONS OF DYNAMICAL PHYSICS

4. Newton's Laws of Motion. 4·1 The theoretical difficulties in the way of the application of the philosophic doctrine of relativity have never worried practical scientists. They have started with the working assumptions that in some sense the world is in one euclidean space, that the permanent points in such a space have no individual characteristics recognisable by us, except so far as they are occupied by recognisable material or except in so far as they are defined by assigned spatial relations to points which are thus definitely recognisable, and that according to the purpose in hand either the earth can be assumed to be at rest or else astronomical axes which are defined by the aid of the solar system, of the stars, and of dynamical considerations deduced from Newton's laws of motion.

4·2 Newton's laws* of motion presuppose the notions of mass and force. Mass arises from the conception of a passive quality of a material body, what it is in itself apart from its relation to other bodies; the notion of 'force' is that of an active agency changing the physical circumstances of the body, and in particular its spatial relations to other bodies. It is fairly obvious that mass and force were introduced into science as the outcome of this antithesis between intrinsic quality and agency, although further reflection may somewhat mar the simplicity of this outlook. Mass and

* Cf. Appendix I to this chapter.

force are measurable quantities, and their numerical expressions are dependent on the units chosen. The mass of a body is constant, so long as the body remains composed of the same self-identical material. Velocity, acceleration and force are vector quantities, namely they have direction as well as magnitude. They are thus representable by straight lines drawn from any arbitrary origin.

4·3 These laws of motion are among the foundations of science; and certainly any alteration in them must be such as to produce effects observable only under very exceptional circumstances. But, as is so often the case in science, a scrutiny of their meaning produces many perplexities.

In the first place we can sweep aside one minor difficulty. In our experience, a finite mass of matter occupies a volume and not a point. Evidently therefore the laws should be stated in an integral form, involving at certain points of the exposition greater elaboration of statement. These forms are stated (with somewhat abbreviated explanation) in dynamical treatises.

Secondly, Lorentz's distinction between macroscopic equations and microscopic equations forces itself on us at once, by reason of the molecular nature of matter and the dynamical nature of heat. A body apparently formed of continuous matter with its intrinsic geometrical relations nearly invariable is in fact composed of agitated molecules. The equations of motion for such a body as used by an engineer or an astronomer are, in Lorentz's nomenclature, macroscopic. In such equations even a differential element of volume is to be supposed to be sufficiently large to average out the diverse agitations of the molecules, and to register only the general unbalanced residuum which to ordinary observation is the motion of the body.

The microscopic equations are those which apply to the individual molecules. It is at once evident that a series of such sets of equations is possible, in which the adjacent sets are macroscopic and microscopic relatively to each other. For example, we may penetrate below the molecule to the electrons and the core which compose it, and thus obtain infra-molecular equations. It is purely a question as to whether there are any observed phenomena which in this way receive their interpretation.

The inductive evidence for the validity of Newton's equations of motion, within the experimental limits of accuracy, is obviously much stronger in the case of the macroscopic equations of the engineer and the astronomer than it is in the case of the microscopic equations of the molecule, and very much stronger than in the case of the infra-microscopic equations of the electron. But there is good evidence that even the infra-microscopic equations conform to Newton's laws as a first approximation. The traces of deviation arise when the velocities are not entirely negligible compared to that of light.

4·4   What do we know about masses and about forces? We obtain our knowledge of forces by having some theory about masses, and our knowledge of masses by having some theory about forces. Our theories about masses enable us in certain circumstances to assign the numerical ratios of the masses of the bodies involved; then the observed motions of these bodies will enable us to register (by the use of Newton's laws of motion) the directions and magnitudes of the forces involved, and thence to frame more extended theories as to the laws regulating the production of force. Our theories about the direction and comparative magnitudes of forces and the observed motions of the bodies will enable us to register (by the use of Newton's laws of motion) the comparative magnitudes of masses. The final results are to be found in engineers' pocket-books in tables of physical constants for physicists, and in astronomical tables. The verification is the concordant results of diverse experiments. One essential part of such theories is the judgment of circumstances which are sufficiently analogous to warrant the assumption of the same mass or the same magnitude of force in assigned diverse cases. Namely the theories depend upon the fact of recognition.

4·5 It has been popular to define force as the product of mass and acceleration. The difficulty to be faced with this definition is that the familiar equation of elementary dynamics, namely,

$$mf = P,$$

now becomes $\qquad mf = mf.$

It is not easy to understand how an important science can issue from such premisses. Furthermore the simple balancing of a weight by the tension of the supporting spring receives a very artificial meaning. With equal reason we might start with our theories of force as fundamental, and define mass as force divided by acceleration. Again we should be in equal danger of reducing dynamical equations to such identities as

$$P/f = P/f.$$

Also the permanent mass of a bar of iron receives a very artificial meaning.

5. The Ether. 5·1 The theory of stress between distant bodies, considered as an ultimate fact, was repudiated by Newton himself, but was adopted by some of his immediate successors. In the nineteenth century the belief in action at a distance has steadily lost ground.

There are four definite scientific reasons for the adoption of the opposite theory of the transmission of stress through an intermediate medium which we will call the 'ether.' These reasons are in addition to the somewhat vague philosophic preferences, based on the disconnection involved in spatial and temporal separation. In the first place, the wave theory of light also postulates an ether, and thus brings concurrent testimony to its existence. Secondly, Clerk Maxwell produced the formulae for the stresses in such an ether which, if they exist, would account for gravitational, electrostatic, and magnetic attractions. No theory of the nature of the ether is thereby produced which in any way explains why such stresses exist; and thus their existence is so far just as much a disconnected assumption as that of the direct stresses between distant bodies. Thirdly, Clerk Maxwell's equations of the electromagnetic field presuppose events and physical properties of apparently empty space. Accordingly there must be something, i.e. an ether, in the empty space to which these properties belong. These equations are now recognised as the foundations of the exact science of electromagnetism, and stand on a level with Newton's equations of motion. Thus another testimony is added to the existence of an ether.

Lastly, Clerk Maxwell's identification of light with electromagnetic waves shows that the same ether is required by the apparently diverse optical and electromagnetic phenomena. The objection is removed that fresh properties have to be ascribed to the ether by each of the distinct lines of thought which postulate it.

It will be observed that gravitation stands outside this unification of scientific theory due to Maxwell's work, except so far that we know the stresses in the ether which would produce it.

5·2 The assumption of the existence of an ether at once raises the question as to its laws of motion. Thus in addition to the hierarchy of macroscopic and microscopic equations, there are the equations of motion for ether in otherwise empty space. The *à priori* reasons for believing that Newton's laws of motion apply to the ether are very weak, being in fact nothing more than the inductive extension of laws to cases widely dissimilar from those for which they have been verified. It is however a sound scientific procedure to investigate whether the assumed properties of ether are explicable on the assumption that it is behaving like ordinary matter, if only to obtain suggestions by contrast for the formulation of the laws which do express its physical changes.

The best method of procedure is to assume certain large principles deducible from Newton's laws and to interpret certain electromagnetic vectors as displacements and velocities of the ether. In this way Larmor has been successful in deducing Maxwell's equations from the principle of least action after making the necessary assumptions. In this he is only following a long series of previous scientists who during the nineteenth century devoted themselves to the explanation of optical and electromagnetic phenomena. His work completes a century of very notable achievement in this field.

5·3 But it may be doubted whether this procedure is not an inversion of the more fundamental line of thought. It will have been noted that Newton's equations, or any equivalent principles which are substituted for them, are in a sense merely blank forms. They require to be supplemented by hypotheses respecting the nature of the stresses, of the masses, and of the motions, before there can be any possibility of their application. Thus by the time that Newton's equations of motion are applied to the explication of etherial events there is a large accumulation of hypotheses respecting things of which we know very little. What in fact we do know about the ether is summed up in Maxwell's equations, or in recent adaptations of his equations such as those due to Lorentz. The discovery of electromagnetic mass and electromagnetic momentum suggests that, for the ether at least, we gain simpler conceptions of the facts by taking Maxwel 's equations, or the Lorentz-Maxwell equations, as fundamental. Such equations would then be the ultimate microscopic equations, at least in the present stage of science, and Newton's equations become macroscopic equations which apply in certain definite circumstances to etherial aggregates. Such a procedure does not prejudge the debated theory of the purely electromagnetic origin of mass.

5·4 The modern theory of the molecule is destructive of the obviousness of the prejudgment in favour of the traditional concepts of ultimate material at an instant. Consider a molecule of iron. It is composed of a central core of positive electricity surrounded by annular clusters of electrons, composed of negative electricity and rotating round the core. No single characteristic property of iron as such can be manifested at an instant. Instantaneously there is simply a distribution of electricity and Maxwell's equations to express our expectations. But iron is not an expectation or even a recollection. It is a fact; and this fact, which is iron, is what happens during a period of time. Iron and a biological organism are on a level in requiring time for functioning. There is no such thing as iron at an instant; to be iron is a character of an event. Every physical constant respecting iron which appears in scientific tables is the register of such a character. What is ultimate in iron, according to the traditional theory, is instantaneous distributions of electricity; and this ultimateness is simply ascribed by reason of a metaphysical theory, and by no reason of observation.

5·5 In truth, when we have once admitted the hierarchy of macroscopic and microscopic equations, the traditional concept is lost. For it is the macroscopic equations which express the facts of immediate observation, and these equations essentially express the integral characters of events. But this hierarchy is necessitated by every concept of modern physics—the molecular theory of matter, the dynamical theory of heat, the wave theory of light, the electromagnetic theory of molecules, the electromagnetic theory of mass.

6. *Maxwell's Equations**. 6·1 A discussion of Maxwell's equations would constitute a treatise on electromagnetism. But they exemplify some general considerations on physical laws.

These equations (expressed for an axis-system a) involve for each point of space and each instant of time the vector quantities $(F_a,\ G_a,\ H_a)$, $(L_a,\ M_a,\ N_a)$ and

* Cf. Appendix II to this chapter.

(u_a, v_a, w_a), namely the electric and magnetic 'forces' and the velocity of the charge of electricity. Now a vector involves direction; and direction is not concerned with what is merely at that point. It is impossible to define direction without reference to the rest of space; namely, it involves some relation to the whole of space.

Again the equations involve the spatial differential operators $\frac{\partial}{\partial x_a}, \frac{\partial}{\partial y_a}, \frac{\partial}{\partial z_a}$, which enter through the symbols curl_a and div_a; and they also involve the temporal differential operator $\frac{\partial}{\partial t_a}$. The differential coefficients thus produced essentially express properties in the neighbourhood of the point (x_a, y_a, z_a) and of the time t_a, and not merely properties at (x_a, y_a, z_a, t_a). For a differential coefficient is a limit, and the limit of a function at a given value of its argument expresses a property of the aggregate of the values of the function corresponding to the aggregate of the values of the argument in the neighbourhood of the given value.

This is essentially the same argument as that expressed above in 1·2 for the particular case of motion. Namely, we cannot express the facts of nature as an aggregate of individual facts at points and at instants.

6·2   In the Lorentz-Maxwell equations [cf. Appendix II] there is no reference to the motion of the ether. The velocity (u_a, v_a, w_a) which appears in them is the velocity of the electric charge. What then are the equations of motion of the ether? Before we puzzle over this question, a preliminary doubt arises. Does the ether move?

Certainly, if science is to be based on the data included in the Lorentz-Maxwell equations, even if the equations be modified, the motion of the ether does not enter into experience. Accordingly Lorentz assumes a stagnant ether: that is to say, an ether with no motion, which is simply the ultimate entity of which the vectors (F_a, G_a, H_a) and (L_a, M_a, N_a) express properties. Such an ether has certainly a very shadowy existence; and yet we cannot assume that it moves, merely for the sake of giving it something to do.

6·3 The ultimate facts contemplated in Maxwell's equations are the occurrences of ρ_a (the volume-density of the charge), (u_a, v_a, w_a), (F_a, G_a, H_a), and (L_a, M_a, N_a) at the space-time points in the neighbourhood surrounding the space-time point (x_a, y_a, z_a, t_a). But this is merely to say that the ultimate facts contemplated by Maxwell's equations are certain events which are occurring throughout all space. The material called ether is merely the outcome of a metaphysical craving. The continuity of nature is the continuity of events; and the doctrine of transmission should be construed as a doctrine of the coextensiveness of events with space and time and of their reciprocal interaction. In this sense an ether can be admitted; but, in view of the existing implication of the term, clearness is gained by a distinction of phraseology. We shall term the traditional ether an 'ether of material' or a 'material ether,' and shall employ the term 'ether of events' to express the assumption of this enquiry, which may be loosely stated as being 'that something is going on everywhere and always.' It is our purpose to express accurately the relations between these events so far as they are disclosed by our perceptual experience, and in particular to consider those relations from which the essential concepts of Time, Space, and persistent material are derived. Thus primarily we must not conceive of events as in a given Time, a given Space, and consisting of changes in given persistent material. Time, Space, and Material are adjuncts of events. On the old theory of relativity, Time and Space are relations between materials; on our theory they are relations between events.

APPENDIX I TO CHAPTER II

NEWTON'S LAWS OF MOTION

Let $(O_a X_a Y_a Z_a)$ as in the accompanying figure be rectangular axes at rest; let $(\dot{x}_{ap}, \dot{y}_{ap}, \dot{z}_{ap})$ be the velocity of a material particle p of mass m at (x_a, y_a, z_a) relative to these axes, and let $(\ddot{x}_{ap}, \ddot{y}_{ap}, \ddot{z}_{ap})$ be the acceleration of the same particle. Also let (X_{ap}, Y_{ap}, Z_{ap}) be the force on the particle p. The first two of Newton's laws can be compressed into the equations

$$m\ddot{x}_{ap} = X_{ap}, \quad m\ddot{y}_{ap} = Y_{ap}, \quad m\ddot{z}_{ap} = Z_{ap}.\ldots\ldots\ldots(1)$$

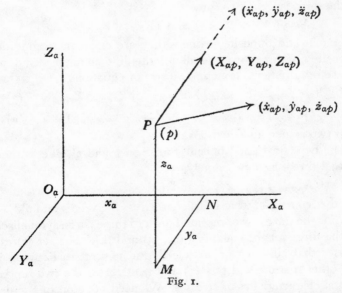

Fig. 1.

It is unnecessary to trace the elementary consequences of these equations.

The third law of motion considers a fundamental characteristic of force and is founded on the sound principle that all agency is nothing else than relations between those entities which are among the ultimate data of science. The law is, Action and reaction are equal and opposite. This means that there must be particles p', p'', p''', etc. to whose agency (X_{ap}, Y_{ap}, Z_{ap}) are due, and that we can write

$$\left.\begin{aligned} X_{ap} &= X_{app'} + X_{app''} + \ldots \\ Y_{ap} &= Y_{app'} + Y_{app''} + \ldots \\ Z_{ap} &= Z_{app'} + Z_{app''} + \ldots \end{aligned}\right\}, \quad \ldots\ldots\ldots\ldots(2)$$

where $(X_{app'}, Y_{app'}, Z_{app'})$ is due to p' alone, $(X_{app''}, Y_{app''}, Z_{app''})$ to p'' alone, and so on.

Furthermore let the particle p' be at (x_a', y_a', z_a') and $(\ddot{x}'_{ap'}, \ddot{y}'_{ap'}, \ddot{z}'_{ap'})$ be the acceleration of p'. Also let $(X_{ap'}, Y_{ap'}, Z_{ap'})$ be the force on p'; and let $X_{ap'p}$, $Y_{ap'p}$, etc. have meanings for p' analogous to those which $X_{app'}$, etc. have for p. Then according to the third law the two forces

$$(X_{app'}, Y_{app'}, Z_{app'}) \text{ and } (X_{ap'p}, Y_{ap'p}, Z_{ap'p})$$

are equal and opposite, namely they are equal in magnitude, opposite in direction, and along the line joining p and p'. These requirements issue in two sets of equations

$$X_{app'} + X_{ap'p} = 0, \; Y_{app'} + Y_{ap'p} = 0, \; Z_{app'} + Z_{ap'p} = 0 \quad (3)$$

and $\quad (y_a Z_{app'} - z_a Y_{app'}) + (y_a' Z_{ap'p} - z_a' Y_{ap'p}) = 0, \; \ldots(4)$

with two analogous equations.

The two equal and opposite forces on p and p', due to their mutual direct agency, namely,

$$(X_{app'}, Y_{app'}, Z_{app'}) \text{ and } (X_{ap'p}, Y_{ap'p}, Z_{ap'p}),$$

together constitute what is called a 'stress between p and p'.'

Thus the third law of motion falls into three parts, symbolised by the three sets of equations (2), (3) and (4). The set (2) expresses that all force on matter is due to stresses between it and other matter; and sets (3) and (4) express the two fundamental characteristics of stresses. We need not stop to enquire whether the short verbal expression of the law logically expresses these three properties. This is a minor point of exposition dependent on the context in which this formulation of the law is found.

APPENDIX II TO CHAPTER II

MAXWELL'S EQUATIONS

It will be convenient to state these equations in the slightly modified form which is due to Lorentz. Space is referred to the fixed rectangular axis system a, as in subarticle $6 \cdot 1$. It will be necessary to explain a few small points of nomenclature and notation.

A vector is a directed physical quantity; for example, the electric force at a point is a vector. This example also shows that we have to conceive vectors which have analogous significations at different points of space. Such a vector is the electric force which may have a distinct magnitude and direction at each point of space, but expresses at all points one definite physical fact. Such a vector will be a function of its position, that is to say, of the coordinates of the point (x_a, y_a, z_a) of which it is that characteristic vector.

Let (X_a, Y_a, Z_a) be any such vector. Then X_a and Y_a and Z_a are each of them functions of (x_a, y_a, z_a) and also of the time t_a, i.e. they are functions of x_a, y_a, z_a, t_a. We shall assume that our physical quantities are differentiable, except possibly at exceptional points.

Let $q(X_a, Y_a, Z_a)$ stand for (qX_a, qY_a, qZ_a), and analogously

$$\frac{\partial}{\partial t}(X_a, Y_a, Z_a) \text{ for } \left(\frac{\partial X_a}{\partial t_a}, \frac{\partial Y_a}{\partial t_a}, \frac{\partial Z_a}{\partial t_a}\right);$$

also $\text{div}_a (X_a, Y_a, Z_a)$ for $\dfrac{\partial X_a}{\partial x_a} + \dfrac{\partial Y_a}{\partial y_a} + \dfrac{\partial Z_a}{\partial z_a}$,

and $\text{curl}_a (X_a, Y_a, Z_a)$ for the vector

$$\left(\frac{\partial Z_a}{\partial y_a} - \frac{\partial Y_a}{\partial z_a}, \frac{\partial X_a}{\partial z_a} - \frac{\partial Z_a}{\partial x_a}, \frac{\partial Y_a}{\partial x_a} - \frac{\partial X_a}{\partial y_a}\right).$$

Finally if (X_a', Y_a', Z_a') be another vector at the same point. then $[(X_a', Y_a', Z_a') . (X_a, Y_a, Z_a)]$

stands for what is called the 'vector product' of the two vectors, namely the vector

$$(Y_a'Z_a - Z_a'Y_a, \ Z_a'X_a - X_a'Z_a, \ X_a'Y_a - Y_a'X_a).$$

It is evident that $\text{curl}_a(X_a, \ Y_a, \ Z_a)$ can be expressed in the symbolic form

$$\left[\left(\frac{\partial}{\partial x_a}, \ \frac{\partial}{\partial y_a}, \ \frac{\partial}{\partial z_a}\right) . (X_a, \ Y_a, \ Z_a)\right].$$

The vector equation

$$(X_a, \ Y_a, \ Z_a) = (X_a', \ Y_a', \ Z_a')$$

is an abbreviation of the three equations

$$X_a = X_a', \ Y_a = Y_a', \ Z_a = Z_a'.$$

Let (F_a, G_a, H_a) be the electric force at (x_a, y_a, z_a, t_a), and let (L_a, M_a, N_a) be the magnetic force at the same point and time. Also let ρ_a be the volume density of the electric charge and (u_a, v_a, w_a) its velocity; and let (P_a, Q_a, R_a) be the ponderomotive force: all equally at (x_a, y_a, z_a, t_a). Finally let c be the velocity of light *in vacuo*.

Then Lorentz's form of Maxwell's equations is

$$\text{div}_a(F_a, \ G_a, \ H_a) = \rho_a, \dots\dots\dots\dots\dots(1)$$

$$\text{div}_a(L_a, \ M_a, \ N_a) = 0, \ \dots\dots\dots\dots\dots(2)$$

$$\text{curl}_a(L_a, M_a, N_a) = \frac{1}{c}\left\{\frac{\partial}{\partial t_a}(F_a, G_a, H_a) + \rho_a(u_a, v_a, w_a)\right\}, \quad (3)$$

$$\text{curl}_a(F_a, G_a, H_a) = -\frac{1}{c}\frac{\partial}{\partial t_a}(L_a, M_a, N_a), \dots\dots\dots(4)$$

$$(P_a, Q_a, R_a) = (F_a, G_a, H_a) + \frac{1}{c}[(u_a, v_a, w_a) . (L_a, M_a, N_a)]. \ (5)$$

It will be noted that each of the vector equations (3), (4), (5) stands for three ordinary equations, so that there are eleven equations in the five formulae.

CHAPTER III

SCIENTIFIC RELATIVITY

7. *Consentient Sets.* 7·1 A traveller in a railway carriage sees a fixed point of the carriage. The wayside stationmaster knows that the traveller has been in fact observing a track of points reaching from London to Manchester. The stationmaster notes his station as fixed in the earth. A being in the sun conceives the station as exhibiting a track in space round the sun, and the railway carriage as marking out yet another track. Thus if space be nothing but relations between material bodies, points as simple entities disappear. For a point according to one type of observation is a track of points according to another type. Galileo and the Inquisition are only in error in the single affirmation in which they both agreed, namely that absolute position is a physical fact—the sun for Galileo and the earth for the Inquisition.

7·2 Thus each rigid body defines its own space, with its own points, its own lines, and its own surfaces. Two bodies may agree in their spaces; namely, what is a point for either may be a point for both. Also if a third body agrees with either, it will agree with both. The complete set of bodies, actual or hypothetical, which agree in their space-formation will be called a 'consentient' set.

The relation of a 'dissentient' body to the space of a consentient set is that of motion through it. The dissentient body will itself belong to another consentient set. Every body of this second set will have a motion in the space of the first set which has the same general spatial characteristics as every other body of the second consentient set; namely (in technical language) it will at any instant be a screw motion with the same axis, the same pitch and the same intensity—in short the same screw-motion for all bodies of the second set. Thus we will speak of the motion of one consentient set in the space of another consentient set. For example such a motion may be translation without rotation, and the translation may be uniform or accelerated.

7·3 Now observers in both consentient sets agree as to what is happening. From different standpoints in nature they both live through the same events, which in their entirety are all that there is in nature. The traveller and the stationmaster both agree as to the existence of a certain event—for the traveller it is the passage of the station past the train, and for the stationmaster it is the passage of the train past the station. The two sets of observers merely diverge in setting the same events in different frameworks of space and (according to the modern doctrine) also of time.

This spatio-temporal framework is not an arbitrary convention. Classification is merely an indication of characteristics which are already there. For example, botanical classification by stamens and pistils and petals applies to flowers, but not to men. Thus the space of the consentient set is a fact of nature; the traveller with the set only discovers it.

8. *Kinematic Relations.* 8·1 The theory of relative motion is the comparison of the motion of a consentient set β in the space of a consentient set α with the motion of α in the space of β. This involves a preliminary

comparison of the space of α with the space of β. Such a comparison can only be made by reference to events which are facts common to all observers, thus showing the fundamental character of events in the formation of space and time. The ideally simple event is one indefinitely restricted both in spatial and in temporal extension, namely the instantaneous point. We will use the term 'event-particle' in the sense of 'instantaneous point-event.' The exact meaning of the ideal restriction in extension of an event-particle will be investigated in Part III; here we will assume that the concept has a determinate signification.

8·2 An event-particle occupies instantaneously a certain point in the space of α and a certain point in the space of β. Thus instantaneously there is a certain correlation between the points of the space of α and the points of the space of β. Also if the particle has the character of material at rest at the point in the space of α, this material-particle has a certain velocity in the space of β; and if it be material at rest at the point in the space of β, the material-particle has a certain velocity in the space of α. The direction in β-space of the velocity due to rest in the correlated α-point is said to be opposite to the direction in α-space of the velocity due to rest in the correlated β-point. Also with congruent units of space and of time, the measures of the velocities are numerically equal. The consequences of these fundamental facts are investigated in Part III. The relation of the α-space to the β-space which is expressed by the velocities at points in α-space due to rest in the points of β-space and by the opposite velocities in β-space due to rest in the points of α-space is called the 'kinematic relation' between the two consentient sets, or between the two spaces.

8·3 The simplest form of this kinematic relation between a pair of consentient sets is when the motion of either set in the space of the other is a uniform translation without acceleration and without rotation. Such a kinematic relation will be called 'simple.' If a consentient group α has a simple kinematic relation to each of two consentient sets, β and γ, then β and γ have a simple kinematic relation to each other. In technical logical language a simple kinematic relation is symmetrical and transitive.

The whole group of consentient sets with simple kinematic relations to any one consentient set, including that set itself, is called a 'simple' group of consentient sets.

The kinematic relation is called 'translatory' when the relative motion does not involve rotation; namely, it is a translation but not necessarily uniform.

8·4 The fact that the relational theory of space involves that each consentient set has its own space with its own peculiar points is ignored in the traditional presentation of physical science. The reason is that the absolute theory of space is not really abandoned, and the relative motion, which is all that can be observed, is treated as the differential effect of two absolute motions.

8·5 In the enunciation of Newton's Laws of Motion, the velocities and accelerations of particles must be supposed to refer to the space of some given consentient set. Evidently the acceleration of a particle is the same in all the spaces of a simple group of consentient sets— at least this has hitherto been the unquestioned assumption. Recently this assumption has been questioned and does not hold in the new theory of relativity. Its axiomatic obviousness only arises from the covert assumption of absolute space. In the new theory Newton's equations themselves require some slight modification which need not be considered at this stage of the discussion.

In either form, their traditional form or their modified form, Newton's equations single out one and only one simple group of consentient sets, and require that the motions of matter be referred to the space of any one of these sets. If the proper group be chosen the third law of action and reaction holds. But if the laws hold for one simple group, they cannot hold for any other such group. For the apparent forces on particles cannot then be analysed into reciprocal stresses in the space of any set not a member of the original simple group.

Let the simple group for which the laws do hold be called the 'Newtonian' group.

8·6 Then, for example, if a consentient set α have a non-uniform translatory kinematic relation to members of the Newtonian group, the particles of the material universe would, when their motions are referred to the α-space, appear to be acted on by forces parallel to a fixed direction, in the same sense along that direction, and proportional to the mass of the particle acted on. Such an assemblage of forces cannot be expressed as an assemblage of reciprocal stresses between particles. Again if a consentient set β have a non-translatory kinematic relation to the members of the Newtonian group, then, when motion is referred to the β-space, 'centrifugal' and 'composite centrifugal' forces on particles make their appearance; and these forces cannot be reduced to stresses.

8·7 The physical consequences of this result are best seen by taking a particular case. The earth is rotating and its parts are held together by their mutual gravitational attractions. The result is that the figure bulges at the equator; and, after allowing for the deficiencies of our observational knowledge, the results of theory and experiment are in fair agreement.

The dynamical theory of this investigation does not depend on the existence of any material body other than the earth. Suppose that the rest of the material universe were annihilated, or at least any part of it which is visible to our eye-sight. Why not? For after all there is a very small volume of visible matter compared to the amount of space available for it. So there is no reason to assume anything very essential in the existence of a few planets and a few thousand stars. We are left with the earth rotating. But rotating relatively to what? For on the relational theory it would seem to be the mutual relations of the earth's parts which constitute space. And yet the dynamical theory of the bulge does not refer to any body other than the earth, and so is not affected by the catastrophe of annihilation. It has been asserted that after all the fixed stars are essential, and that it is the rotation relatively to them which produces the bulge. But surely this ascription of the centrifugal force on the earth's surface to the influence of Sirius is the last refuge of a theory in distress. The point is that the physical properties, size, and distance of Sirius do not seem to matter. The more natural deduction (on the theory of Newtonian relativity) is to look on the result as evidence that the theory of any empty space is an essential impossibility. Accordingly the absoluteness of direction is evidence for the existence of the material ether. This result only reinforces a conclusion which has already been reached on other grounds. Thus space expresses mutual relations of the parts of the ether, as well as of the parts of the earth.

9. *Motion through the Ether.* 9·1 The existence of the material ether should discriminate between the consentient sets of the Newtonian group. For one such set will be at rest relatively to the ether, and the remaining sets will be moving through it with definite velocities. It becomes a problem to discover phenomena dependent on such velocities.

Can any phenomena be detected which are unequivocally due to a quasi-absolute motion of the earth through the ether? For this purpose we must put aside phenomena which depend on the differential velocities of two bodies of matter, e.g. the earth and a planet, or a star. For such phenomena are evidently primarily due to the relative velocity of the two bodies to each other, and the velocities relatively to the ether only arise as a hypothetical intermediate explanatory analysis. We require phenomena concerned solely with the earth, which are modified by the earth's motion through the ether without reference to any other matter. We have already concluded that the bulging of the earth at the equator is one such required instance, unless indeed (with Newton) we assume absolute space.

9·2 The effects on the observed light due to the relative motions of the emitting body and the receiving body are various and depend in part on the specific nature of the assumed disturbances which constitute light. Some of these effects have been observed, for example, aberration and the effect on the spectrum due to the motion of the emitting body in the line of sight. Aberration is the apparent change in the direction of the luminous body due to the motion of the receiving body. The motion of the luminous body in the line of sight should alter the wave length of the emitted light due to molecular vibrations of given periodicity. In other words, it should alter the quality of the light due to such vibrations. These are the effects which have been observed, but they are of the type which we put aside as not relevant to our purpose owing to the fact that the observed effect ultimately depends merely on the relative motion of the emitting and receiving bodies.

9·3 There are effects on interference fringes which we should expect to be due to the motion of the earth. In six months the velocity of the earth in its orbit is reversed. So that such effects as the earth's motion produces in the interference fringes of a certain purely terrestrial apparatus at one time can be compared with the corresponding effects in the same apparatus which it produces after the lapse of six months and—as the experiments have been carried out—the differences should have been easily discernible. No such differences have been observed. The effects, which are thus sought for, depend on no special theory of the nature of the luminous disturbance in the ether. They should result from the simple fact of the wave disturbance, and the magnitude of its velocity relatively to the apparatus.

It will be observed that the difficulty which arises from the absence of this predicted effect does not discriminate in any way between the philosophic theories of absolute or of relative space. The effect should arise from the motion of the earth relatively to the ether, and there is such relative motion whichever of the alternative spatial theories be adopted.

9·4 Electromagnetic phenomena are also implicated in the theory of relative motion. Maxwell's equations of the electromagnetic field hold in respect to these phenomena an analogous position to that occupied by Newton's equations of motion for the explanation of the motion of matter. They differ from Newton's equations very essentially in their relation to the principle of relativity. Newton's equations single out no special member of the Newtonian group to which they specially apply. They are invariant for the spatio-temporal transformations from one such set to another within the Newtonian group.

But Maxwell's electromagnetic equations are not thus invariant for the Newtonian group. The result is that they must be construed as referring to one particular consentient set of this group. It is natural to suppose that this particular assumption arises from the fact that the equations refer to the physical properties of a stagnant ether; and that accordingly the consentient set presupposed in the equations is the consentient set of this ether. The ether is identified with the ether whose wave disturbances constitute light; and furthermore there are practically conclusive reasons for believing light to be merely electromagnetic disturbances which are governed by Maxwell's equations.

The motion of the earth through the ether affects other electromagnetic phenomena in addition to those known to us as light. Such effects, as also in the case of light, would be very small and difficult to observe. But the effect on the capacity of a condenser of the six-monthly reversal of the earth's velocity should under proper conditions be observable. This is known as Trouton's experiment. Again, as in the analogous case of light, no such effect has been observed.

9·5 The explanation [the Fitzgerald-Lorentz hypothesis] of these failures to observe expected effects has been given, that matter as it moves through ether automatically readjusts its shape so that its lengths in the direction of motion are altered in a definite ratio dependent on its velocity. The null results of the experiments are thus completely accounted for, and the material ether evades the most obvious method of testing its existence. If matter is thus strained by its passage through ether, some effect on its optical properties due to the strains might be anticipated. Such effects have been sought for, but not observed. Accordingly with the assumption of an ether of material the negative results of the various experiments are explained by an *ad hoc* hypothesis which appears to be related to no other phenomena in nature.

9·6 There is another way in which the motion of matter may be balanced (so to speak) against the velocity of light. Fizeau experimented on the passage of light through translucent moving matter, and obtained results which Fresnel accounted for by multiplying the refractive index of the moving medium by a coefficient dependent on its velocity. This is Fresnel's famous 'coefficient of drag.' He accounted for this coefficient by assuming that as the material medium in its advance sucks in the ether, it condenses it in a proportion dependent on the velocity. It might be expected that any theory of the relations of matter to ether, either an ether of material or an ether of events, would explain also this coefficient of drag.

10. Formulae for Relative Motion. 10·1 In transforming the equations of motion from the space of one member of the Newtonian group to the space of another member of that group, it must be remembered that the facts which are common to the two standpoints are the events, and that the ideally simple analysis exhibits events as dissected into collections of event-particles. Thus if α and β be the two consentient sets, the points of the α-space are distinct from the points of the β-space, but the same event-particle e is at the point P_α at the time T_α in the α-space and is at the point P_β at the time T_β in the β-space.

With the covert assumption of absolute space which is habitual in the traditional outlook, it is tacitly assumed that P_α and P_β are the same point and that there is a common time and common measurement of time which are the same for all consentient sets. The first assumption is evidently very badly founded and cannot easily be reconciled to the nominal scientific creed; the second assumption seems to embody a deeply rooted experience. The corresponding formulae of transformation which connect the measurements of space, velocity, and acceleration in the α-system for space and time with the corresponding measurements in the β-system certainly are those suggested by common sense and in their results they agree very closely with the result of careful observation. These formulae are the ordinary formulae of dynamical treatises. For such transformations the Newtonian equations are invariant within the Newtonian group.

10·2 But, as we have seen, this invariance, with these

formulae for transformation, does not extend to Maxwell's equations for the electromagnetic field. The
conclusion is that—still assuming these formulae for
transformation—Maxwell's equations apply to the
electromagnetic field as referred to one particular
consentient set of the Newtonian group. It is natural
to suppose that this set should be that one with respect
to which the stagnant ether is at rest. Namely, stating
the same fact conversely, the stagnant ether defines
this consentient set. There would be no difficulty about
this conclusion except for the speculative character of
the material ether, and the failure to detect the evidences
of the earth's motion through it. This consentient set
defined by the ether would for all practical purposes
define absolute space.

10·3   There are however other formulae of transformation from the space and time measurements of
set α to the space and time measurements of set β for
which Maxwell's equations are invariant. These formulae were discovered first by Larmor for uncharged
regions of the field and later by Lorentz for the general
case of regions charged or uncharged. Larmor and
Lorentz treated their discovery from its formal mathematical side. This aspect of it is important. It enables
us, when we thoroughly understand the sequence of
events in one electromagnetic field, to deduce innumerable other electromagnetic fields which will be understood equally well. All mathematicians will appreciate
what an advance in knowledge this constitutes.

But Lorentz also pointed out that if these formulae for
transformation could be looked on as the true formulae
for transformation from one set to another of the
Newtonian group, then all the unsuccessful experiments

to detect the earth's motion through the ether could be explained. Namely, the results of the experiments are such as theory would predict.

10·4 The general reason for this conclusion was given by Einstein in a theorem of the highest importance. He proved that the Lorentzian formulae of transformation from one consentient set to another of the Newtonian group—from set α to set β—are the necessary and sufficient conditions that motion with the one particular velocity c (the velocity of light *in vacuo*) in one of the sets, α or β, should also appear as motion with the same magnitude c in the other set, β or α. The phenomena of aberration will be preserved owing to the relation between the directions of the velocity expressing the movements in α-space and β-space respectively. This preservation of the magnitude of a special velocity (however directed) cannot arise with the traditional formulae for relativity. It practically means that waves or other influences advancing with velocity c as referred to the space of any consentient set of the Newtonian group will also advance with the same velocity c as referred to the space of any other such set.

10·5 At first sight the two formulae for transformation, namely the traditional formulae and the Lorentzian formulae, appear to be very different. We notice however that, if α and β be the two consentient sets and if $V_{\alpha\beta}$ be the velocity of β in the α-space and of α in the β-space, the differences between the two formulae all depend upon the square of the ratio of $V_{\alpha\beta}$ to c, where c is the velocity of light *in vacuo*, and are negligible in proportion to the smallness of this number. For ordinary motions, even planetary motions, this ratio is extremely small and its square is smaller still. Accordingly the differences between the two formulae would not be perceptible under ordinary circumstances. In fact the effect of the difference would only be perceived in those experiments, already discussed, whose results have been in entire agreement with the Lorentzian formulae.

The conclusion at once evokes the suggestion that the Lorentzian formulae are the true formulae for transformation from the space and time relations of a consentient set α to those of a consentient set β, both sets belonging to the Newtonian group. We may suppose that, owing to bluntness of perception, mankind has remained satisfied with the Newtonian formulae which are a simplified version of the true Lorentzian relations. This is the conclusion that Einstein has urged.

Fig. 2.

10·6 These Lorentzian formulae for transformation involve two consequences which are paradoxical if we covertly assume absolute space and absolute time. Let α and β be two consentient sets of the Newtonian group. Let an event-particle P happen at the point P_α in the α-space and at the point P_β in the β-space, and let another event-particle Q happen at points Q_α and Q_β in the two spaces respectively. Then according to the traditional scientific outlook, P_α and P_β are not discriminated from each other; and similarly for Q_α and Q_β. Thus evidently the distance $P_\alpha Q_\alpha$ is (on this theory) equal to $P_\beta Q_\beta$, because in fact they are symbols for the same distance. But if the true distinction between the α-space and the β-space is kept in mind, including the fact that the points in the two spaces are radically distinct, the equality of the distances $P_\alpha Q_\alpha$ and $P_\beta Q_\beta$ is not so obvious. According to the Lorentzian formulae such corresponding distances in the two spaces will not in general be equal.

The second consequence of the Lorentzian formulae involves a more deeply rooted paradox which concerns our notions of time. If the two event-particles P and Q happen simultaneously when referred to the points P_α and Q_α in the α-space, they will in general not happen simultaneously when referred to the points P_β and Q_β in the β-space. This result of the Lorentzian formulae contradicts the assumption of one absolute time, and makes the time-system depend on the consentient set which is adopted as the standard of reference. Thus there is an α-time as well as an α-space, and a β-time as well as a β-space.

10·7 The explanation of the similarities and differences between spaces and times derived from different consentient sets of the Newtonian group, and of the fact of there being a Newtonian group at all, will be derived in Parts II and III of this enquiry from a consideration of the general characteristics of our perceptive knowledge of nature, which is our whole knowledge of nature. In seeking such an explanation one principle may be laid down. Time and space are among the fundamental physical facts yielded by our knowledge of the external world. We cannot rest content with any theory of them which simply takes mathematical equations involving four variables (x, y, z, t) and interprets (x, y, z) as space coordinates and t as a measure of time, merely on the ground that some physical law is thereby expressed. This is not an interpretation of what we *mean* by space and time. What we mean are physical facts expressible in terms of immediate perceptions; and it is incumbent on us to produce the perceptions of those facts as the meanings of our terms.

Einstein has interpreted the Lorentzian formulae in terms of what we will term the 'message' theory, discussed in the next chapter.

APPENDIX TO CHAPTER III

Let α and β be two consentient sets of the Newtonian group. Let $(O_\alpha X_\alpha Y_\alpha Z_\alpha)$ be the rectangular axis system in the space of α, and $(O_\beta' X_\beta' Y_\beta' Z_\beta')$ be the rectangular axis system in the space of β.

First consider the traditional theory of relativity. Then the time-system is independent of the consentient set of reference.

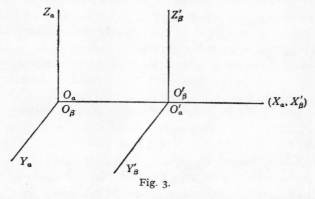

Fig. 3.

At the time t let the event-particle which instantaneously happens at the point O_α in the space of α happen at O_β in the space of β, and let the event-particle which happens at O_β' in the space of β happen at O_α' in the space of α. Let the axis $O_\alpha X_\alpha$ be in the direction of the motion of β in the α-space, and the axis $O_\beta' X_\beta'$ be in the direction reversed of the motion of α in the β-space. Also let O_β' be so chosen that O_α' lies on $O_\alpha X_\alpha$. Then the event-particles at the instant t which happen on $O_\alpha X_\alpha$ are the event-particles which happen at the instant t on $O_\beta' X_\beta'$. Also we choose $O_\beta' Y_\beta'$ and $O_\beta' Z_\beta'$ so that the event-particles which happen at time t on $O_\beta' Y_\beta'$ and $O_\beta' Z_\beta'$ respectively happen on straight lines in the α-space which are parallel to $O_\alpha Y_\alpha$ and $O_\alpha Z_\alpha$. Let $V_{\alpha\beta}$ be the velocity of β in α-space and

$V_{\beta a}$ be the velocity of α in β-space. Then (with a suitable origin of time)

$$\left.\begin{array}{l} V_{\alpha\beta} + V_{\beta a} = 0, \\ O_a O_a' = O_\beta O_{\beta}', \\ x_a = x_\beta + V_{\alpha\beta} t, \ y_a = y_\beta, \ z_a = z_\beta. \end{array}\right\} \quad\ldots\ldots\ldots\ldots(\text{I})$$

These are the 'Newtonian' formulae for relative motion.

Secondly consider the Lorentzian [or 'electromagnetic'] theory of relativity. The two time-systems for reference to α and for reference to β respectively are not identical. Let t_a be the measure of the lapse of time in the α-system, and t_β be the measure of the lapse of time in the β-system. The distinction between the two time-systems is embodied in the fact that event-particles which happen simultaneously at time t_a in α-space do not happen simultaneously throughout space β. Thus supposing that an event-particle happens at (x_a, y_a, z_a, t_a) in α-space and α-time and at $(x_\beta, y_\beta, z_\beta, t_\beta)$ in β-space and β-time, we seek for the formulae which are to replace equations (I) of the Newtonian theory.

As before let $O_a X_a$ lie in the direction of the motion of β in α, and $O_{\beta}' X_{\beta}'$ in the reverse direction of the motion of α in β. Also let O_a' lie on $O_a X_a$, so that event-particles which happen on $O_a X_a$ also happen on $O_{\beta}' X_{\beta}'$. One connection between the two time-systems is secured by the rule that event-particles which happen simultaneously at points in α-space on a plane perpendicular to $O_a X_a$ also happen simultaneously at points in β-space on a plane perpendicular to $O_{\beta}' X_{\beta}'$. Accordingly the quasi-parallelism of $O_a Y_a$ to $O_{\beta}' Y_{\beta}'$, and of $O_a Z_a$ to $O_{\beta}' Z_{\beta}'$, is defined and secured in the same way as for Newtonian relativity.

The same meaning as above will be given to $V_{\alpha\beta}$ and $V_{\beta a}$; also c is the fundamental velocity which is the velocity of light *in vacuo*. Then we define

$$\Omega_{\alpha\beta} = \Omega_{\beta a} = (1 - V_{\alpha\beta}{}^2/c^2)^{-\frac{1}{2}}. \quad\ldots\ldots\ldots\ldots(2)$$

The formulae for transformation are

$$\left.\begin{array}{l} V_{\alpha\beta} + V_{\beta a} = 0, \\ t_\beta = \Omega_{\alpha\beta}(t_a - V_{\alpha\beta} x_a/c^2), \\ x_\beta = \Omega_{\alpha\beta}(x_a - V_{\alpha\beta} t_a), \\ y_\beta = y_a, \ z_\beta = z_a. \end{array}\right\} \quad\ldots\ldots\ldots\ldots(3)$$

These formulae are symmetrical as between α and β, so that

$$\left.\begin{aligned}t_\alpha &= \Omega_{\beta a}\,(t_\beta - V_{\beta a} x_\beta / c^2),\\x_\alpha &= \Omega_{\beta a}\,(x_\beta - V_{\beta a} t_\beta).\end{aligned}\right\} \quad \dots\dots\dots(4)$$

It is evident that when $V_{a\beta}/c$ is small,

$$\Omega_{a\beta} = \Omega_{\beta a} = \mathrm{I},$$

and when x_α and x_β are not too large

$$t_\beta = t_\alpha,$$
$$x_\beta = x_\alpha - V_{a\beta} t_\alpha.$$

Thus the formulae reduce to the Newtonian type.

Let $\dot{x}_a,\ \dot{y}_a,\ \dot{z}_a$ stand for $\dfrac{dx_a}{dt_a}$, etc., and $\dot{x}_\beta,\ \dot{y}_\beta,\ \dot{z}_\beta$ for $\dfrac{dx_\beta}{dt_\beta}$, etc.

Then it follows immediately from the preceding formulae that

$$\left.\begin{aligned}\dot{x}_\beta &= (\dot{x}_a - V_{a\beta})\Big/\left(\mathrm{I} - \frac{V_{a\beta}\dot{x}_a}{c^2}\right),\\[4pt]\dot{y}_\beta &= \Omega_{a\beta}^{-1}\dot{y}_a\Big/\left(\mathrm{I} - \frac{V_{a\beta}\dot{x}_a}{c^2}\right),\\[4pt]\dot{z}_\beta &= \Omega_{a\beta}^{-1}\dot{z}_a\Big/\left(\mathrm{I} - \frac{V_{a\beta}\dot{x}_a}{c^2}\right).\end{aligned}\right\} \quad \dots\dots\dots(5)$$

With the notation of Appendix II to Chapter II, the formulae of transformation for Maxwell's equations are

$$\left.\begin{aligned}F_\beta &= F_a,\\[2pt]G_\beta &= \Omega_{a\beta}\left(G_a - \frac{V_{a\beta}}{c}N_a\right),\\[2pt]H_\beta &= \Omega_{a\beta}\left(H_a + \frac{V_{a\beta}}{c}M_a\right);\end{aligned}\right\} \quad \dots\dots\dots(6)$$

and

$$\left.\begin{aligned}L_\beta &= L_a,\\[2pt]M_\beta &= \Omega_{a\beta}\left(M_a + \frac{V_{a\beta}}{c}H_a\right),\\[2pt]N_\beta &= \Omega_{a\beta}\left(N_a - \frac{V_{a\beta}}{c}G_a\right);\end{aligned}\right\} \quad \dots\dots\dots(7)$$

and

$$\rho_\beta = \Omega_{a\beta}\rho_a\left(\mathrm{I} - \frac{V_{a\beta}u_a}{c^2}\right), \quad \dots\dots\dots(8)$$

where (u_a, v_a, w_a) is the velocity of the charge at (x_a, y_a, z_a) at the time t_a.

Also it immediately follows from formulae (5) that

$$\Omega_{a\beta}{}^2(\dot{x}_\beta{}^2 + \dot{y}_\beta{}^2 + \dot{z}_\beta{}^2 - c^2) = (\dot{x}_a{}^2 + \dot{y}_a{}^2 + \dot{z}_a{}^2 - c^2)\Big/\Big(1 - \frac{V_{a\beta}\dot{x}_a}{c^2}\Big)^2.$$

Hence

$$(\dot{x}_\beta{}^2 + \dot{y}_\beta{}^2 + \dot{z}_\beta{}^2 - c^2) \text{ and } (\dot{x}_a{}^2 + \dot{y}_a{}^2 + \dot{z}_a{}^2 - c^2)$$

vanish together. This proves Einstein's theorem on the invariance of the velocity c, so far as concerns the sufficiency of the Lorentzian formulae to produce that result.

CHAPTER IV

CONGRUENCE

11. Simultaneity. 11·1 Einstein analysed the ideas of time-order and of simultaneity. Primarily (according to his analysis) time-order only refers to the succession of events at a given place. Accordingly each given place has its own time-order. But these time-orders are not independent in the system of nature, and their correlation is known to us by means of physical measurements. Now ultimately all physical measurement depends upon coincidence in time and place. If P_1 and P_2 be two places, the time-orders O_1 and O_2 which belong to P_1 and P_2 are correlated by observations of coincidences at P_1 and at P_2 respectively.

Thus, confining ourselves to the two places P_1 and P_2, there are two distinct processes of correlating the time-order of events throughout the universe, namely by a series of observations of coincidences at P_1 based on time-order O_1 and by a series of observations of coincidences at P_2 based on time-order O_2. These two processes are distinct and will only agree by some accident of special circumstance.

11·2 What are the observations at P_1 which will assign to an event at P_2 a position in the time-order O_1? Suppose some message—a wave disturbance, for example—starts from P_1 when event e_1 happens at P_1, reaches P_2 when event e_2 happens at P_2, and is immediately reflected so as to return to P_1 when event e_1'' happens at P_1. Now according to the method of time-measurement for O_1, there is an event

e_1' which happens at mid-time between e_1 and e_1''. Then, when certain conditions have been fulfilled, the event e_2 at P_2 is defined as simultaneous with the event e_1' at P_1 according to the method of correlation appropriate to place P_1. In this way a time-order of events at P_2 is derived solely from observation of coincidence at P_1 and is based solely on the fundamental time-order O_1 at P_1. Thus the time-order at P_1 is extended as a time-order for all events at all places.

11·3 There are questions which require elucidation before this definition can be understood. What is a place? We have chosen a vague term on purpose, so as to postpone its consideration until now. A place can only be marked by phenomena capable of recognition, for example the continued appearance of a material body. Thus we must construe P_1 and P_2 to be the names of material bodies, or of persistent sets of circumstances which will serve the same purpose. In general P_1 and P_2 will be in relative motion with respect to each other.

What of the message which passes from P_1 to P_2 and back to P_1? Its transmission must be uniform. Suppose the message travels with velocity c, that is, with the velocity of light *in vacuo*. Then, assuming the electromagnetic formulae for relativity, this velocity relative to P_1 is independent (so far as its magnitude is concerned) of the velocity which we ascribe to P_1 through space.

11·4 Thus our recording body P_1 can be any body at rest in some consentient set of the Newtonian group, and we reckon motion as relative to the space of this set. We send our message with the velocity of light *in vacuo*. Then, according to the local time-order O_1

at P_1, the event e_2 at P_2 is simultaneous with the event $e_1{}'$ at P_1. This definition of simultaneity in the local time-order at P_1 is independent of any assumption of absolute rest for P_1, provided that the electromagnetic formulae for relativity are adopted. The local time-order at P_1 is also in complete agreement with the local time-order at any body Q_1 which is rigidly connected with P_1, i.e. which belongs to the same consentient set.

11·5 The reason why the velocity of light has been adopted as the standard velocity in the definition of simultaneity is because the negative results of the experiments to determine the earth's motion require that this velocity, which is the 'c' of Maxwell's equations, should have this property. Also light signals are after all our only way of detecting distant events.

Certainly, once granting the idea of time-order being a local affair connected with a specific body P_1, the acceptance of the electromagnetic formula connecting t_α and t_β is a slight affair. There is no presumption against it, once granting the conception of diverse time-orders which had not hitherto been thought of.

11·6 But there are certain objections to the acceptance of Einstein's definition of simultaneity, the 'signal-theory' as we will call it. In the first place light signals are very important elements in our lives, but still we cannot but feel that the signal-theory somewhat exaggerates their position. The very meaning of simultaneity is made to depend on them. There are blind people and dark cloudy nights, and neither blind people nor people in the dark are deficient in a sense of simultaneity. They know quite well what it means to bark both their shins at the same instant. In fact the determination of simultaneity in this way is never made, and if it could be made would not be accurate; for we live in air and not *in vacuo*.

Also there are other physical messages from place to place; there is the transmission of material bodies, the transmission of sound, the transmission of waves and ripples on the surface of water, the transmission of nerve excitation through the body, and innumerable other forms which enter into habitual experience. The transmission of light is only one form among many.

Furthermore local time does not concern one material particle only. The same definition of simultaneity holds throughout the whole space of a consentient set in the Newtonian group. The message theory does not account for the consentience in time-reckoning which characterises a consentient set, nor does it account for the fundamental position of the Newtonian group.

12. Congruence and Recognition. 12·1 Again the theory that measurement is essentially coincidence requires severe qualification. For if it were true only coincident things, coincident both in time and space, could be equal, yet measurement can only be of the slightest importance in so far as some other element not coincidence enters into it.

Let us take a simple example. Two footrules are placed together and are found to coincide. Then at the moment of coincidence they are equal in length. But what is the use of that information? We want to use one rule to-morrow in London and the other rule a week hence in Manchester, and to know that the stuffs which they measure are of equal length. Now we know that, provided they are made of certain sorts of material (luckily, materials easy to procure) and treated with certain precautions (luckily, precautions easy to observe), the footrules will not have altered their lengths to any extent which can be detected. But that means a direct judgment of constancy. Without such a judgment in some form or other, measurement becomes trivial.

12·2 It may be objected that whenever the footrules are brought together, or when stuffs measured by them are brought together, the coincidences will be observed; and that this is all we need for the importance of measurement.

But the coincidences will not be observed unless the circumstances of the various experiments are sufficiently uniform. The stuffs must be under the same tension or at the same temperatures as on previous occasions. Sooner or later and somehow or other a judgment of constancy, that is, of the preservation of property, is required. Ultimately this judgment reposes upon direct common sense; namely, obviously the footrule is of good stiff material and has not perceptibly changed amid slight differences of circumstance. The coincidences which can easily be obtained between lengths of elastic thread inspire no such beliefs, because evidently the thread has been stretched.

12·3 Again, in Einstein's own example, there is the direct judgment of the uniformity of conditions for the uniform transmission of light. Thus any ordinary event among the fixed stars does not affect this uniformity for the transmission from the sun to the earth. Apart from such presuppositions, so obvious that they do not enter into consciousness, the whole theory collapses.

12·4 These judgments of constancy are based on an

immediate comparison of circumstances at different times and at different places. Such judgments are not infallible and are capable of being tested under certain circumstances. For example it may be judged that two footrules would coincide if they were brought together; and this experiment can be made, and the judgment tested.

The rejection of an immediate judgment of constancy is no paradox. There are differences between any distinct sets of circumstances, and it is always possible that these differences cut deeper than we have perceived so as to produce unsuspected divergences of properties.

But a judgment of constancy is recognition, and recognition is the source of all our natural knowledge. Accordingly though isolated judgments may be rejected, it is essential that a rational consideration of nature should assume the truth of the greater part of such judgments and should issue in theories which embody them.

12·5 This recognition of congruity between distinct circumstances has no especial connection with coincidence and extends far beyond the mere judgments of time and space. Thus judgments of the matching of colours can be made without coincidence by most people to some slight extent, and by some people with surprising accuracy. It may be urged that only in the case of judgments of spatial and temporal coincidence can great accuracy be obtained. This may be true; but complete accuracy is never obtained, and the ideal of accuracy shows that the meaning is not derived from the measurement. Our recognitions are the ultimate facts of nature for science, and the whole scientific theory is nothing else than an attempt to systematise our knowledge of the circumstances in which such recognitions will occur. The theory of congruence is one branch of the more general theory of recognitions. Another branch is the theory of objects which is considered in the next part of this enquiry.

PART II

THE DATA OF SCIENCE

CHAPTER V

THE NATURAL ELEMENTS

13. The Diversification of Nature. 13·1 Our perceptual knowledge of nature consists in the breaking up of a whole which is the subject matter of perceptual experience, or is the given presentation which is experience—or however else we prefer to describe the ultimate experienced fact. This whole is discriminated as being a complex of related entities, each entity having determinate qualities and relations and being a subject concerning which our perceptions, either directly or indirectly, afford definite information. This process of breaking up the subject matter of experience into a complex of entities will be called the ' diversification of nature.'

13·2 This diversification of nature is performed in different ways, according to different procedures which yield different analyses of nature into component entities. It is not merely that one mode of diversification of nature is incomplete and leaves out some entities which another mode supplies. The entities which are yielded by different modes of diversification are radically different; and it is the neglect of this distinction between the entities of complexes produced by different modes of diversification which has produced so much confusion in the principles of natural knowledge.

There are an indefinite number of types of entity disclosed in this diversification. An attempt in this enquiry to trace the subtlety of nature would only blur the main argument. Accordingly we confine attention to five modes of diversification which are chiefly important in scientific theory. These types of entities are: (i) events, (ii) percipient objects, (iii) sense-objects, (iv) perceptual objects, (v) scientific objects. These are five radically distinct types of entities yielded by five distinct procedures; and their only common quality as entities is that they are all alike subjects yielded for our knowledge by our perceptions of nature.

13·3 The entities which are the product of any one mode of diversification of nature will be called elements, or aspects, of nature; each such entity is one natural element. Thus each mode of diversification produces natural elements of a type peculiar to itself.

One mode of diversification is not necessarily more abstract than another. Objects can be looked on as qualities of events, and events as relations between objects, or—more usefully—we can drop the metaphysical and difficult notion of inherent qualities and consider the elements of different types as bearing to each other relations.

There are accordingly two main genera of relations to be distinguished, namely 'homogeneous' relations which relate among themselves natural elements of the same type, and 'heterogeneous' relations which relate natural elements of different types.

13·4 Another way of considering the diversification of nature is to emphasise primarily the relations between natural elements. Thus those elements are what is perceived in nature as thus related. In other words the relations are treated as fundamental and the natural elements are introduced as in their capacity of relata. But of course this is merely another mode of expression, since relations and relata imply each other.

14. Events. 14·1 Events are the relata of the fundamental homogeneous relation of 'extension.' Every event extends over other events which are parts of itself, and every event is extended over by other events of which it is part. The externality of nature is the outcome of this relation of extension. Two events are mutually external, or are 'separate,' if there is no event which is part of both. Time and space both spring from the relation of extension. Their derivation will be considered in detail in subsequent parts of this enquiry. It follows that time and space express relations between events. Other natural elements which are not events are only in time and space derivatively, namely, by reason of their relations to events. Great confusion has been caused to the philosophy of science by this neglect of the derivative nature of the spatial and temporal relations of objects of various types.

14·2 The relation of extension exhibits events as actual—as matters of fact—by means of its properties which issue in spatial relations; and it exhibits events as involving the becomingness of nature—its passage or creative advance—by means of its properties which issue in temporal relations. Thus events are essentially elements of actuality and elements of becomingness. An actual event is thus divested of all possibility. It is what does become in nature. It can never happen again; for essentially it is just itself, there and then. An event is just what it is, and is just how it is related and it is nothing else. Any event, however similar, with different relations is another event. There is no element of hypothesis in any actual event. There are imaginary events—or, rather, imaginations of events—but there is nothing actual about such events, except so far as imagination is actual. Time and space, which are entirely actual and devoid of any tincture of possibility, are to be sought for among the relations of events.

14·3 Events never change. Nature develops, in the sense that an event e becomes part of an event e' which includes (i.e. extends over) e and also extends into the futurity beyond e. Thus in a sense the event e does change, namely, in its relations to the events which were not and which become actual in the creative advance of nature. The change of an event e, in this meaning of the term 'change,' will be called the 'passage' of e; and the word 'change' will not be used in this sense. Thus we say that events pass but do not change. The passage of an event is its passing into some other event which is not it.

An event in passing becomes part of larger events; and thus the passage of events is extension in the making. The terms 'the past,' 'the present,' and 'the future' refer to events. The irrevocableness of the past is the unchangeability of events. An event is what it is, when it is, and where it is. Externality and extension are the marks of events; an event is there and not here [or, here and not there], it is then and not now [or, now and not then], it is part of certain wholes and is a whole extending over certain parts.

15. Objects. 15·1 Objects enter into experience by recognition and without recognition experience would divulge no objects. Objects convey the permanences recognised in events, and are recognised as self-identical

amid different circumstances; that is to say, the same object is recognised as related to diverse events. Thus the self-identical object maintains itself amid the flux of events: it is there and then, and it is here and now; and the 'it' which has its being there and here, then and now, is without equivocation the same subject for thought in the various judgments which are made upon it.

15·2 The change of an object is the diverse relationships of the same object to diverse events. The object is permanent, because (strictly speaking) it is without time and space; and its change is merely the variety of its relations to the various events which are passing in time and in space. This passage of events in time and space is merely the exhibition of the relations of extension which events bear to each other, combined with the directional factor in time which expresses that ultimate becomingness which is the creative advance of nature. These extensional relations of events are analysed in later parts of this enquiry. But here we merely make clear that change in objects is no derogation from their permanence, and expresses their relation to the passage of events; whereas events are neither permanent nor do they change. Events (in a sense) are space and time, namely, space and time are abstractions from events. But objects are only derivatively in space and time by reason of their relations to events.

15·3 The ways in which events and objects enter into experience are distinct. Events are lived through, they extend around us. They are the medium within which our physical experience develops, or, rather, they are themselves the development of that experience. The facts of life are the events of life.

Objects enter into experience by way of the intellectuality of recognition. This does not mean that every object must have been known before; for in that case there never could have been a first knowledge. We must rid our imagination of the fallacious concept of the present as instantaneous. It is a duration, or stretch of time; and the primary recognition of an object consists of the recognition of its permanence amid the partial events of the duration which is present. Its recognition is carried beyond the present by means of recollection and memory.

Rational thought which is the comparison of event with event would be intrinsically impossible without objects.

15·4 Objects and events are only waveringly discriminated in common thought. Whatever is purely matter of fact is an event. Whenever the concept of possibility can apply to a natural element, that element is an object. Namely, objects have the possibility of recurrence in experience: we can conceive imaginary circumstances in which a real object might occur. The essence of an object does not depend on its relations, which are external to its being. It has in fact certain relations to other natural elements; but it might (being the same object) have had other relations. In other words, its self-identity is not wholly dependent on its relations. But an event is just what it is, and is just how it is related; and it is nothing else.

Thus objects lack the fixedness of relations which events possess, and thus time and space could never be a direct expression of their essential relations. Two objects have (by the mediation of events) all the mutual space relations which they do have throughout their existence, and might have many which they do not have. Thus two objects, being what they are, have no necessary temporal and spatial relations which are essential to their individualities.

15·5 The chief confusion between objects and events is conveyed in the prejudice that an object can only be in one place at a time. That is a fundamental property of events; and whenever that property appears axiomatic as holding of some physical entity, that entity is an event. It must be remembered however that ordinary thought wavers confusedly between events and objects. It is the misplacement of this axiom from events to objects which has wrecked the theory of natural objects.

15·6 It is an error to ascribe parts to objects, where 'part' here means spatial or temporal part. The erroneousness of such ascription immediately follows from the premiss that primarily an object is not in space or in time. The absence of temporal parts of objects is a commonplace of thought. No one thinks that part of a stone is at one time and another part of the stone is at another time. The same stone is at both times, in the sense in which the stone is existing at those times (if it be existing). But spatial parts are in a different category, and it is natural to think of various parts of a stone, simultaneously existing. Such a conception confuses the stone as an object with the event which exhibits the actual relations of the stone within nature. It is indeed very natural to ascribe spatial parts to a stone, for the reason that a stone is an instance of a perceptual object. These objects are the objects of common life, and it is very difficult precisely to discern such an object in the events with which it has its most obvious relations. The struggle to make precise the concept of these objects either forces us back to the sense-objects or forward to the scientific objects. The difficulty is chiefly one of making thought clear. That there is a perception of an object with self-identity, is shown by the common usage of mankind. Indeed these perceptual objects forced upon mankind—and seemingly also on animals, unless it be those of the lowest type—their knowledge of the objectified character of nature. But the confusion of the object, which is a unity, with the events, which have parts, is always imminent. In biological organisms the character of the organism as an object is more clear.

15·7 The fundamental rule is that events have parts and that—except in a derivative sense, from their relations to events—objects have no parts. On the other hand the same object can be found in different parts of space and time, and this cannot hold for events. Thus the identity of an object may be an important physical fact, while the identity of an event is essentially a trivial logical necessity. Thus the prisoner in the dock may be the man who did the deed. But the deed lies in the irrevocable past; only the allegation of it is before the court and perhaps (in some countries) a reconstitution of the crime. Essentially the very deed itself is never there.

15·8 The continuity of nature is to be found in events, the atomic properties of nature reside in objects. The continuous ether is the whole complex of events; and the atoms and molecules are scientific objects, which are entities of essentially different type to the events forming the ether.

15·9 This contrast in the ways we perceive events

and objects deserves a distinction in nomenclature. Accordingly, for want of better terms, we shall say that we 'apprehend' an event and 'recognise' an object. To apprehend an event is to be aware of its passage as happening in that nature, which we each of us know as though it were common to all percipients. It is unnecessary for the purposes of science to consider the difficult metaphysical question of this community of nature to all. It is sufficient that, for the awareness of each, it is as though it were common to all, and that science is a body of doctrine true for this quasi-common nature which is the subject for the experience of each percipient; namely, science is true for each percipient.

To recognise an object is to be aware of it in its specific relations to definite events in nature. Thus we refer the object to some events as its 'situations,' we connect it with other events as the locus from which it is being perceived, and we connect it with other events as conditions for such perception of it as in such situations from such a locus of percipience.

Accordingly in these (arbitrary) senses of the words we apprehend nature as continuous and we recognise it as atomic.

CHAPTER VI

EVENTS

16. Apprehension of Events. 16·1 It is the purpose of this chapter to summarise the leading characteristics of our knowledge of nature as diversified into a complex of events.

Perception is an awareness of events, or happenings, forming a partially discerned complex within the background of a simultaneous whole of nature. This awareness is definitely related to one event, or group of events, within the discerned complex. This event is called the percipient event. The simultaneity of the whole of nature comprising the discerned events is the special relation of that background of nature to the percipient event. This background is that complete event which is the whole of nature simultaneous with the percipient event, which is itself part of that whole. Such a complete whole of nature is called a 'duration.' A duration (in the sense in which henceforth the word will be used) is not an abstract stretch of time, and to that extent the term 'duration' is misleading. In perception the associated duration is apprehended as an essential element in the awareness, but it is not discriminated into all its parts and qualities. It is the complete subject matter for a discrimination which is only very partially performed.

Thus the whole continuum of nature 'now-present' means one whole event (a duration), rendered definite by the limitation 'now-present' and extending over all events now-present. Namely, the various finite events now-present for an awareness are all parts of one associated duration which is a special type of event. A duration is in a sense unbounded; for it is, within certain limitations, all that there is. It has the property of completeness, limited by the condition 'now-present'; it is a temporal slab of nature.

16·2 This fact of nature as a present-whole is forced on our apprehension by the character of perception. Perceptual awareness is complex. There are the various types of sense-perception, and differences in extensity and in intensity. There are also differences in attention and in consequent clearness of awareness, shading off into a dim knowledge of events barely on the threshold of consciousness. Thus nature, as we know it, is a continuous stream of happening immediately present and partly dissected by our perceptual awareness into separated events with diverse qualities. Within this present stream the perceived is not sharply differentiated from the unperceived; there is always an indefinite 'beyond' of which we feel the presence although we do not discriminate the qualities of the parts. This knowledge of what is beyond discriminating perception is the basis of the scientific doctrine of externality. There is a present-whole of nature of which our detailed knowledge is dim and mediate and inferential, but capable of determination by its congruity with clear immediate perceptual facts.

16·3 The condition 'now-present' specifies a particular duration. It evidently refers to some relation; for 'now' is 'simultaneous with,' and 'present' is 'in the presence of' or 'presented to.' Thus 'now-present' refers to some relation between the duration and something else. This 'something else' is the event 'here-present,' which is the definite connecting link between

individual experienced knowledge and self-sufficient
nature. The essential existence of the event 'here-
present' is the reason why perception is from within
nature and is not an external survey. It is the 'percipient
event.' The percipient event defines its associated
duration, namely its corresponding 'all nature.'

16·4 The 'here' in 'here-present' also refers to the
specific relation between the percipient event and its
associated duration. It means 'here within the duration,' i.e. 'here within the present continuum of nature.'
Thus the relation between an event 'here-present' and
its associated duration embodies in some form the
property of rest in the duration; for otherwise 'here'
would be an equivocation. The relation in any concrete
case may be complex, involving more than one meaning
of 'here,' but the essential character of the relation
is that as we (according to the method of extensive
abstraction) properly diminish the extent of such an
event, the property of 'rest in' the associated duration
becomes more obvious. When an event has the property of being a percipient event unequivocally here
within an associated duration, we shall say that it is
'cogredient' with the duration.

16·5 An event can be cogredient with only one
duration. To have this relation to the duration it must
be temporally present throughout the duration and
exhibit one specific meaning of 'here.' But a duration
can have many events cogredient with it. Namely any
event, which is temporally present throughout that
duration and in relation to an event here-present defines
one specific meaning of 'there,' is an event 'there-
present' which has the same relation of cogredience to
that duration and (to that extent) is (so far) potentially

an event 'here-present' in that duration for some possible act of apprehension. Thus cogredience is a condition for a percipient event yielding unequivocal meanings to 'here' and 'now.'

The relation of cogredience presupposes that the duration extends over the event; but the two relations must not be confounded. In the first place a duration extends over events which are not temporally present throughout it, so that the specification of the duration would not be a complete answer to the question 'When?' as asked of the event. Secondly, the question 'Where?' which means 'Where in the duration?' may not be susceptible of the one definite answer 'There' which is only possible if cogredience holds. The question 'Whither?' which contemplates change in the 'there' of an event, definitely refers to events which are parts of a duration but are not cogredient with it. Cogredience is the relation of absolute position within a duration; we must remember that a duration is a slab of nature and not a mere abstract stretch of time. Cogredience is the relation which generates the consentient sets discussed in Chapter III of Part I. The details of the deduction belong to Part III.

16·6 It is not necessary to assume that there is one event which is the system of all nature throughout all time. For scientific purposes the only unbounded events are durations and these are bounded in their temporal extension.

17. The Constants of Externality. **17·1** The 'constants of externality' are those characteristics of a perceptual experience which it possesses when we assign to it the property of being an observation of the passage of external nature, namely when we apprehend it. A fact which possesses these characteristics, namely these constants of externality, is what we call an 'event.'

A complete enumeration of these constants is not necessary for our purpose; we only need a survey of just those elements in the apprehension of externality from which the concepts of time, space and material arise. In this survey the attitude of mind to be avoided is exhibited in the questions, 'How, being in space, do we know it?' 'How, being in time, do we know it?' and 'How, having material, do we know of it?'

Again we are not considering *à priori* necessities, nor are we appealing to *à priori* principles in proof. We are merely investigating the characteristics which in experience we find belonging to perceived facts when we invest them with externality. The constants of externality are the conditions for nature, and determine the ultimate concepts which are presupposed in science.

17·2 In order to enter upon this investigation from the standpoint of habitual experience, consider the simplest general questions which can be asked of a percipient of some event in nature, 'Which?' 'What?' 'How?' 'When?' 'Where?' 'Whither?' These six questions fall into two sets. The first three invite specification of qualities and discrimination amid alternative entities; the remaining three refer to the spatio-temporal relation of a part to a whole within which in some sense the perceived part is located.

They can be construed as referring to events or to objects. The former way of understanding them is evidently the more fundamental, for our awareness of nature is directly an awareness of events or happenings, which are the ultimate data of natural science. The conditions which determine the nature of events can only be furnished by other events, for there is nothing else in nature. A reference to objects is only a way of specifying the character of an event. It is an error to conceive of objects as causing an event, except in the sense that the characters of antecedent events furnish conditions which determine the natures of subsequent events.

17·3 The ultimate nature of events has been blurred by the confusion which seems to be introduced by its acknowledgment. Events appear as indefinite entities without clear demarcations and with mutual relations of baffling complexity. They seem, so to speak, deficient in thinghood. A lump of matter or a charge of electricity in a position at an instant, retaining its self-identity in other positions at succeeding instants, seems a simple clue for the unravelling of the maze. This may be un-reservedly granted; but our purpose is to exhibit this conception of spatio-temporal material in its true rela-tion to events. When this has been effected, the mechanical rigidity (so to speak) of the traditional views of time, space and material is thereby lost, and the way is opened for such readjustments as the advance of experimental knowledge may suggest.

17·4 The six questions of 17·2 immediately reveal that what is ultimate in nature is a set of determinate things, each with its own relations to other things of the set. To say this is a truism, for thought and judg-ment are impossible without determinate subjects. But the reluctance to abandon a vague indetermination of events has been an implicit reason for the refusal to consider them as the ultimate natural entities.

This demarcation of events is the first difficulty which arises in applying rational thought to experience. In perception no event exhibits definite spatio-temporal limits. A continuity of transition is essential. The definition of an event by assignment of demarcations is an arbitrary act of thought corresponding to no perceptual experience. Thus it is a basal assumption, essential for ratiocination relating to perceptual experience, that there are definite entities which are events; though in practice our experience does not enable us to identify any such subject of thought, as discriminated from analogous subjects slightly more or slightly less.

This assumption must not be construed either as asserting an atomic structure of events, or as a denial of overlapping events. It merely asserts the ideal possibility of perfect definiteness as to what does or does not belong to an event which is the subject of thought, though such definiteness cannot be achieved in human knowledge. It is the claim which is implicit in every advance towards exact observation, namely that there is something definite to be known. The assumption is the first constant of externality, namely the belief that what has been apprehended as a continuum, is a potentially definite complex of entities for knowledge. The assumption is closely allied with the conception of nature as 'given.' This conception is the thought of an event as a thing which 'happened' apart from all theory and as a fact self-sufficient for a knowledge discriminating it alone.

18. Extension. 18·1 The second constant of externality is the relation of extension which holds between events. An event x may 'extend over' an event y,

i.e. in other words y may be part of x. The concepts
of time and of space in the main, though not entirely,
arise from the empirically determined properties of
this relation of extension. It is evident from the universal and uniform application of the spatio-temporal
concepts that they must arise from the utilisation of
the simplest characteristics without which no datum
of knowledge would be recognised as an event belonging
to the order of nature. Extension is a relation of this
type. It is a property so simple that we hardly recognise
it as such—it of course is so. Thus the event which is
the passage of the car is part of the whole life of the
street. Also the passage of a wheel is part of the event
which is the passage of the car. Similarly the event
which is the continued existence of the house extends
over the event which is the continued existence of a
brick of the house, and the existence of the house during
one day extends over its existence during one specified
second of that day.

18·2 Every element of space or of time (as conceived
in science) is an abstract entity formed out of this
relation of extension (in association at certain stages
with the relation of cogredience) by means of a determinate logical procedure (the method of extensive
abstraction). The importance of this procedure depends
on certain properties of extension which are laws of
nature depending on empirical verification. There is,
so far as I know, no reason why they should be so,
except that they are. These laws will be stated in the
succeeding parts so far as is necessary to exemplify the
definitions which are there given and to show that these
definitions really indicate the familiar spatial and temporal entities which are utilised by science in precise

and determinate ways. Many of the laws can be logically proved when the rest are assumed. But the proofs
will not be given here, as our aim is to investigate the
structure of the ideas which we apply to nature and
the fundamental laws of nature which determine their
importance, and not to investigate the deductive science
which issues from them.

18·3 The various elements of time and space are
formed by the repeated applications of the method
of extensive abstraction. It is a method which in
its sphere achieves the same object as does the differential calculus in the region of numerical calcula-
tion, namely it converts a process of approximation
into an instrument of exact thought. The method
is merely the systematisation of the instinctive procedure of habitual experience. The approximate procedure of ordinary life is to seek simplicity of relations
among events by the consideration of events sufficiently
restricted in extension both as to space and as
to time; the events are then 'small enough.' The
procedure of the method of extensive abstraction is
to formulate the law by which the approximation is
achieved and can be indefinitely continued. The complete series is then defined and we have a 'route of
approximation.' These routes of approximation according to the variation of the details of their formation
are the points of instantaneous space (here called
'event-particles'), linear segments (straight or curved)
between event-particles (here called 'routes'), the
moments of time (each of which is all instantaneous
nature), and the volumes incident in moments. Such
elements are the exactly determined concepts on which
the whole fabric of science rests.

18·4 The parts of an event are the set of events (excluding itself) which the given event extends over. It is a mistake to conceive an event as the mere logical sum of its parts. In the first place if we do so, we are necessarily driven back to conceive of more fundamental entities, not events, which would not have the mere abstract logical character which (on this supposition) events would then have. Secondly, the parts of an event are not merely one set of non-overlapping events exhausting the given event. They are the whole complex of events contained in that event; for example, if a be the given event, and a extends over b, and b over c, then a extends over c and both b and c are parts of a. Thus an event has its own substantial unity of being which is not an abstract derivative from logical construction. The physical fact of the concrete unity of an event is the foundation of the continuity of nature from which are derived the precise laws of the mathematical continuity of time and space. Not any two events are in combination just one event, though there will be other events of which both are parts. We recur to this point in Part III, art. 29, when considering the junction of events.

19. Absolute Position. 19·1 The third constant of externality is the fact (already explained) that an event as apprehended is related to a complete whole of nature which extends over it and is the duration associated with the percipient event of that perception.

19·2 The fourth constant of externality is the reference (already explained) of the apprehended event to the percipient event which (when sufficiently restricted in its temporal extension) has a definite station within the associated duration.

19·3 The fifth constant is the above-mentioned fact of the definite station of a percipient event within its duration. Namely, when the specious present is properly limited, there is a definite univocal meaning to the relation 'here within the duration' of the percipient event to the duration.

19·4 Thus the third, fourth, and fifth constants of externality convey its very essence, and without them our perceptual experience appears as a disconnected dream. They embody the reference of an event to a definite—an absolute—spatio-temporal position within a definite whole of nature, which whole is defined and limited by the actual circumstances of the perception. This position, or station, within such a whole is presupposed in the questions, 'When?' 'Where?' 'Whither?'

20. *The Community of Nature.* 20·1 One other constant of externality is required in scientific thought. We will call it the association of events with a 'community of nature.' This sixth constant arises from the fragmentary nature of perceptual knowledge. There are breaks in individual perception, and there are distinct streams of perception corresponding to diverse percipients. For example, as one percipient awakes daily to a fresh perceptual stream, he apprehends the same external nature which can be comprised in one large duration extending over all his days. Again the same nature and the same events are apprehended by diverse percipients; at least, what they apprehend is as though it were the same for all.

20·2 Thus we distinguish between the qualities of events as in individual perception—namely, the sense-data of individuals—and the objective qualities of the actual events within the common nature which is the datum for apprehension. In this assumption of a nature common for all percipients, the immediate knowledge of the individual percipient is entirely his perceptual awareness derived from the bodily event 'now-present here.' But this event occurs as related to the events of antecedent or concurrent nature. Accordingly he is aware of these events as related to his bodily event 'now-present here'; but his knowledge is thus mediate and relative—namely, he only knows other events through the medium of his body and as determined by relations to it. The event here-now, comprising in general the bodily events, is the immediate event conditioning awareness.

20·3 The form that this awareness of nature takes is an awareness of sense-objects now-present, namely qualities situated in the events within the duration associated with the percipient event. Thus the immediate awareness qualifies the events of the specious present. Thus the common nature which is the object of scientific research has to be constructed as an interpretation. This interpretation is liable to error, and involves adjustments. This question is further considered in the next chapter and in Part IV.

21. Characters of Events. 21·1 The characters of events arbitrarily marked out in nature are of baffling complexity. There are two ways of obtaining events of a certain simplicity. In the first place we may consider events cogredient with our present duration. This is in fact to fix attention on a given position in space and to consider what is now going on within it. The spatial relations will be simplified, but (unless we are lucky) the other characters will be very complex. The second method is to consider events whose time-parts show a certain permanence of character. This is in fact to follow the fortunes of objects, and may be termed the natural mode of discriminating the continuous stream of external nature into events. The importance of this mode of discrimination could only be ascertained by experience.

21·2 There is one essential event which each percipient discriminates, namely that event of which each part, contained within each successive duration that assumes for him the character of the duration now-present, correspondingly assumes for him the character of the event here-present. This event is the life of that organism which links the percipient's awareness to external nature.

21·3 The thesis of this chapter can finally be summarised as follows: There is a structure of events and this structure provides the framework of the externality of nature within which objects are located. Any percept which does not find its position within this structure is not for us a percept *of* external nature, though it may find its explanation *from* external events as being derived from them. The character of the structure receives its exposition from the quantitative and qualitative relations of space and time. Space and time are abstractions expressive of certain qualities of the structure. This space-time abstraction is not unique, so that many space-time abstractions are possible, each with its own specific relation to nature. The particular space-time abstraction proper to a particular observant mind depends on the character of the percipient event which is the medium relating that mind to the whole of nature. In a space-time abstraction, time expresses certain qualities of the passage of nature. This passage has also been called the creative advance of nature. But this passage is not adequately expressed by any one time-system. The whole set of time-systems derived from the whole set of space-time abstractions expresses the totality of those properties of the creative advance which are capable of being rendered explicit in thought. Thus no single duration can be completely concrete in the sense of representing a possible whole of all nature without omission. For a duration is essentially related to one space-time system and thus omits those aspects of the passage which find expression in other space-time systems. Accordingly there can be no duration whose bounding moments are the first and last moments of creation.

Objects are entities recognised as appertaining to events; they are the recognita amid events. Events are named after the objects involved in them and according to how they are involved.

CHAPTER VII

OBJECTS

22. Types of Objects. 22·1 We have now to consider natural elements which are objects of various types. There are in fact an indefinite number of such types corresponding to the types of recognisable permanences in nature of various grades of subtlety. It is only necessary here to attempt a rough classification of those which are essential to scientific thought.

The consideration of objects introduces the concepts of 'matter'—or more vaguely, 'material'—of 'transmission' and of 'causation.' These concepts express certain relations of objects to events, but the relations are too complex to be fully expressed in such simple terms.

22·2 The essence of the perception of an object is recognition. There is the primary recognition which is the awareness of permanence within the specious present; there is the indefinite recognition (which we may term 'recollection') which is the awareness of other perceptions of the object as related to other events separate from the specious present, but without any precise designation of the events; and there is the definite recognition (which we may term 'memory') which is an awareness of perception of the object as related to certain other definite events separate from the specious present.

22·3 The awareness of external nature is an awareness of a duration, which is the being of nature throughout the specious present, and of a complex of events, each being part of the present duration. These events fall into two sets. In one set is the percipient event and in the other are the external events whose peculiar property, which has led to their discernment, is that they are the situations of sense-objects.

22·4 The percipient event is discerned as the locus of a recognisable permanence which is the 'percipient object.' This object is the unity of the awareness whose recognition leads to the classification of a train of percipient events as the natural life associated with one consciousness. The discussion of the percipient object leads us beyond the scope of this enquiry. Owing to the temporal duration of the immediate present the self-knowledge of the percipient object is a knowledge of the unity of the consciousness within other parts of the immediate present. Thus, though it is a knowledge of what is immediately present, it is not a knowledge knowing itself.

23. Sense-Objects. 23·1 The sense-object is the simplest permanence which we trace as self-identical in external events. It is some definite sense-datum, such as the colour red of a definite shade. We see redness here and the same redness there, redness then and the same redness now. In other words, we perceive redness in the same relation to various definite events, and it is the same redness which we perceive. Tastes, colours, sounds, and every variety of sensation are objects of this sort.

23·2 There is no apprehension of external events apart from recognitions of sense-objects as related to them, and there is no recognition of sense-objects except as in relation to external events.

In so far as recognition of a sense-object is confined

to primary recognition within the present duration, the sense-object and the event do not clearly disentangle themselves; recollection and memory are the chief agents in producing a clear consciousness of a sense-object. But apart from recollection and memory, any factor, perceived as situated in an external event, which might occur again and which is not a relation between other such factors, is a sense-object. Sense-objects form the ultimate type of perceived objects (other than percipient objects) and do not express any permanence of relatedness between perceived objects of yet more fundamental types.

23·3 A sense-object, such as a particular shade of redness, has a variety of relations to the events of nature. These relations are not explicable in terms of the two-termed relations to which attention is ordinarily confined.

The events which (in addition to the sense-object) enter as terms into such a relation can be classified into three sets (not mutually exclusive), namely (i) percipient events, (ii) events which are 'situations' of the sense-object, (iii) conditioning events.

23·4 A percipient event in the polyadic relation of a sense-object to nature is the percipient event of an awareness which includes the recognition of that sense-object. An event e is a situation of the sense-object for that percipient event when for the associated awareness the sense-object is a quality of e. Now perception involves essentially both a percipient event and an associated duration within which that percipient event is set and with which it is cogredient. A situation of a sense-object in respect to a given percipient event occurs within the associated duration of the percipient event. In fact the content of the awareness derived from a given percipient event is merely the associated duration as extending over a complex of events which are situations of sense-objects of perception and also as extending over the percipient event itself. For example, an astronomer looks through a telescope and sees a new red star burst into existence. He sees redness situated in some event which is happening now and whose spatial relations to other events, though fairly determinate as to direction from him, are very vague as to distance.

23·5 We say that what he really sees is a star coming into being two centuries previously. But the relation of the event 'really seen' to the percipient event and to the redness is an entirely different one from that of the event 'seen' to these same entities. It is only the incurable poverty of language which blurs the distinction.

This distinction between what is 'perceived' and what is 'really perceived' does not solely arise from time differences. For example Alciphron, in Berkeley's dialogue, sees a crimson cloud. Suppose that he had seen the cloud in a mirror. He would have 'seen' crimson as situated in an event behind the mirror, but he would have 'really seen' the cloud behind him.

These examples show that the property of being the situation of a sense-object for a given percipient event is in some respects a trivial property of an event. Yet, in other respects, it is very important; namely, it is important for the consciousness associated with the percipient event. The situations of sense-objects form the whole basis of our knowledge of nature, and the whole structure of natural knowledge is founded on the analysis of their relations.

23·6 The definiteness for human percipients of the situations of sense-objects varies greatly for different types of such objects. The sound of a bell is in the bell, it fills the room, and jars the brain. The feeling a push against a hard rock is associated with the rock as hardness and with the body as effort, where hardness and effort are objects of sense. This duplication of sense-objects is a normal fact in perception, though one of the two associated pair, either the one in the body or the one in a situation separated from the body, is usually faintly perceived and indeterminate as to situation.

23·7 The relationship between a sense-object and nature, so far as it is restricted to one percipient event and one situation, is completed by the conditioning events. The special characters in which they enter into that relationship depend on the particular case under consideration. Conditioning events may be divided into two main classes which are not strictly discriminated from each other. Namely, there are the events which are 'passive' conditions and the events which are 'active' conditions. An event which is an active condition is a cause of the occurrence of the sense-object in its situation for the percipient event; at least, it can be so termed in one of the many meanings of the word 'cause.' Also space and time are presupposed as the setting within which the particular events occur. But space and time are merely expressive of the relations of extension among the whole ether of events. Thus this presupposition of space and time really calls in all events of all nature as passive conditions for that particular perception of the sense-object. The laws of nature express the characters of the active conditioning events and of the percipient events, which issue in the recognition of a definite sense-object in an assigned situation.

23·8 The discovery of laws of nature depends on the fact that in general certain simple types of character of active conditioning events repeat themselves. These are the normal causes of the recognitions of sense-objects. But there are abnormal causes and part of the task in the analysis of natural law is to understand how the abnormal causes are consistent with those laws. For example, the normal cause of the sight of a colour in a situation (near by) is the rectilinear transmission of light (during the specious present) from the situation to the percipient event through intervening events. The introduction of a mirror introduces abnormality. This is an abnormality of a minor sort. An example of major abnormality is when there is no transmission of light at all. The excessive consumption of alcohol produces delirium and illusions of sight. In this example the active conditioning events are of a totally different character from those involving the transmission of light. The perception is a delusion in the sense that it suggests the normal conditioning events instead of the abnormal conditioning events which have actually occurred. Abnormal conditioning events are in no way necessarily associated with error. For example, recollection and memory are perceptions with abnormal conditioning events; and indeed in any abnormal circumstances error only arises when the circumstances are not recognised for what they are.

23·9 Whereas the situations of a sense-object are always simultaneous with the associated percipient event, the active conditioning events are in general antecedent to it. These active conditioning events in general are divisible into two classes not very clearly separated, namely the generating events and the transmitting events. This classification is especially possible in the case of perception under normal circumstances.

24. Perceptual Objects. 24·1 Perceptual objects are the ordinary objects of common experience—chairs, tables, stones, trees. They have been termed 'permanent possibilities of sensation.' These objects are —at least for human beings—the most insistent of all natural objects. They are the 'things' which we see, touch, taste, and hear. The fact of the existence of such objects is among the greatest of all laws of nature, ranking with those from which space and time emerge.

A perceptual object is recognised as an association of sense-objects in the same situation. The permanence of the association is the object which is recognised. It is not the case however that sense-objects are only perceived as associated in perceptual objects. There is always a perception of sense-objects—some sounds, for instance—not so associated. Furthermore, a sense-object associated in a perceptual object is perceived both as itself and as 'conveying' the perceptual object. For example, we see both the horse and the colour of the horse, but what we see (in the strict sense of the term) is simply colour in a situation.

24·2 This property of 'conveying' an object is fundamental in the recognition of perceptual objects. It is the chief example of abnormal perception of sense-objects. It is already well-known in the theory of art-criticism, as is evidenced in such phrases as 'tactile-values' or again in such simple phrases as 'painting water so that it looks wet.'

The conveyance of a perceptual object by a sense-object is not primarily a judgment. It is a sensuous perception of sense-objects, definite as to situation but not very determinate as to exact character. Judgments quickly supervene and form an important ingredient of what may be termed 'completed recognition.' These judgments will be called 'perceptual judgments.'

24·3 Thus in the completed recognition of a perceptual object we discern (i) the primary recognition of one or more sense-objects in the same situation, (ii) the conveyance of other sense-objects by these primary recognitions, and (iii) the perceptual judgment as to the character of the perceptual object which in its turn influences the character of the sense-objects conveyed.

The content of the perceptual judgment is (i) that an analogous association of sense-objects, with 'legal' modifications and in the same situation as that actually apprehended, is recognisable from other percipient events, and (ii) that the event which is the common situation of these associations of sense-objects, recognised or recognisable, is an active condition for these recognitions.

24·4 The situation of a perceptual object is what we call the 'generating' event among the active conditions for the associated sense-objects, provided that the perceptual judgment is correct.

If the perceptual judgment is false, the perceptual object as perceived is a delusion.

The situation of a non-delusive perceptual object is independent of any particular percipient event.

24·5 Amid the development of events the same non-delusive perceptual object may be perceived in a developed situation, again with 'legal' modifications of the association of perceived sense-objects. The verbal analysis of what constitutes a legal modification of the association without breach of the essence of the observed permanence would be impossibly complex in each particular case; but the judgment as to what is allowable in modification is immediate in practice, apart from exceptional cases.

A non-delusive perceptual object will be called a 'physical object.'

It is an essential characteristic of a physical object that its situation is an active condition for its perception. For this reason the object itself is often named as the cause. But the object is only derivatively the cause by its relation to its situation. Primarily a cause is always an event, namely, an active condition.

24·6 The apprehension of an event as the situation of a physical object is our most complete perception of the character of an event. It represents a fundamental perception of a primary law of nature. It is solely by means of physical objects that our knowledge of events as active conditions is obtained, whether as generating conditions or as transmitting conditions. For example, the mirror is recognised as a physical object and its situation is the generating condition for that association of sense-objects; but its situation is also a transmitting condition for the sense-objects and delusive perceptual objects which are perceived as images behind it. Again, the prism is a physical object and its situation is a transmitting condition for the sense-object which is the spectrum.

So far as it is directly perceived in its various situations, a physical object is a group of associations of sense-objects, each association being perceived or perceivable by a percipient object with an appropriate

percipient event as its locus. But the object is more than
the logical group; it is the recognisable permanent
character of its various situations.

24·7 In spite of their insistence in perception these
physical objects are infected with an incurable vagueness which had led speculative physics practically to
cut them out of its scheme of fundamental entities. In
the first place this vagueness arises from the unique
situation of such an object within any small duration.
The result is that the object is confounded with the
event which is its situation. But a situation is prolonged
in time, and a temporal part of that event is not the
event itself. Now the object during ten seconds is
not part of the object during one of these seconds. The
object is always wholly itself during ten seconds or
during one second. It is this train of thought which
led to the introduction of the durationless instant of
time as a fundamental fact, thus fatally confusing the
philosophy of science. The error arose from not discriminating the object from its situation. The train of
events which is the situation of the object through a
prolonged stretch of time is not the unique object; it
is the set of events with which the object has its unique
association. The difficulty of this point of view arises
from the fact that a temporal succession of events, each
very similar to the others, ceases to mark for us the
time-flux in comparison with the rhythmic changes of
our bodies. The result is that in perceiving an unchanging cliff the recognition of permanence, i.e. of
the object, overwhelms all other perception, the flux
of events becoming a vague background owing to the
absence of their demarcation in our perceptual experience.

24·8 The essential unity of the object amid the spatial parts of its situation is more difficult to grasp. The derivation of space and time by the method of extensive analysis, as explained in Part III of this enquiry, exhibits the essential identity of extension in time and extension in space. Thus the reasons for denying temporal parts of an object are also reasons for denying to it spatial parts. Again, it is true that the leg of a chair occupies part of the space which is occupied by the chair. But in appealing to space we are appealing to relations between events. What we are saying is, that the situation of the leg of the chair is part of the situation of the chair. This fact only makes the leg to be part of the chair in a mediate derivative sense, by way of their relations to their situations. But the leg is one object with a recognisable permanence of association, and the chair is another, with recognisable permanence of association distinct from that of the leg, and their situations in all circumstances have certain definite relations to each other expressible* in temporal and spatial terms.

24·9 The second reason for the vagueness of physical objects is the impossibility of submitting the group of associations, forming the object, to any process of determination with a progressive approximation to precision. A physical object is one of those entities of ordinary experience which refuse to be pressed into the service of science by way of a progressive exactness of determination. Consider for example a definite object such as a certain woollen sock. It wears thin, but it remains the same object; it is darned, and remains the same object; finally after successive repairs no part of the

* Cf. Chapters XIV and XV of Part IV.

original wool is left, but it is the same sock. The truth is that each time we affirm the self-identity of this object we are construing the group of associations, which we recognise, in a more and more attenuated sense. The object which is both the sock at the end and the sock at the beginning is a very attenuated complex type of permanence, which would not be what we meant by the sock merely at the beginning of its career or as perceived merely at the end of its career. By insisting on the continued identity of the sock, we are in fact continually juggling with what we mean by the sock, always retaining the most complete associations which we can trace through the whole continuous series of events forming the successive situations of the sock. The physical object 'works' perfectly for the ordinary usage of life, and is thus fully justified for that purpose in the eyes of the pragmatic philosopher.

24·91 But these objects do represent essential facts of nature; sometimes, as it may seem to us, trivial facts not worth disentangling from the events which are their situations, sometimes useful facts. But their essential character is exemplified when we reach biological facts. A living organism exhibits a certain unity of being which is merely the exhibition of the enhanced importance of the unity of the physical object.

25. Scientific Objects. 25·1 The various types of scientific objects arise from the determination of the characters of the active conditioning events which are essential factors in the recognition of sense-objects.

The perceptual judgment which is present in the completed recognition of physical objects introduces the notion of hypothetical perceptions by percipient objects, located for an indefinite number of hypothetical percipient events. In other words, it is a judgment on the events of the universe as being favourable active conditions for the perception of the physical object, granting the correspondingly favourable percipient events. There are an indefinite number of such percipient events, actual or imaginary. The characters of events as active conditions are to be inferred from their adjustment to these innumerable possibilities of perception of each physical object.

25·2 Also in another way physical objects are the links connecting nature as perceived with nature as conditioning its own perception. Physical objects are often termed the causes of the perception of sense-objects, other than the sense-objects which are among their own components. For example, the telescope is the cause of the astronomer's seeing the star. But a physical object is a cause only in an indirect mediate sense. The fact of the telescope being in the right position at the right time was an active condition for the astronomer's sight of the star. Now this fact is an event which is a 'situation' of the telescope. Thus in our experience the situations of physical objects are discovered to be active conditions for the perception of sense-objects. In this way a knowledge of the characters of events, in so far as they are active conditions, can be observed and inferred; and the passage from perception to causation is effected.

25·3 At once the question arises, In what terms are the characters of the conditioning events to be expressed? The unanimous answer has been, that the expression is to be in terms of 'matter,' or—allowing for the more subtle ether—in terms of 'material.' In so deciding none of the distinctions made above have been consistently held in view. The result has been the persistent lapses into confusion which have been exhibited in a brief abstract in the first part of this enquiry.

Matter has been classified into the various kinds of matter which are the chemical substances; thence the atomic theory of matter has been established; and thence some form of electromagnetic theory of molecules is emerging. It is in the last degree unlikely that the present form of this theory will represent its final stage. All novel theories emerge with a childlike simplicity which they ultimately shed. But, apart from specific details, it can as little be doubted that in its main concepts the theory is true.

25·4 We will accordingly pass by the elaborate task of tracking down and interpreting intermediate stages of scientific concepts—important though they are—and pass at once to the consideration of molecules and electrons. The characters of events in their capacity of active conditioning events for sense-objects are expressed by their relations to scientific objects. Scientific objects are not directly perceived, they are inferred by reason of their capacity to express these characters, namely, they express how it is that events are conditions. In other words they express the causal characters of events.

25·5 At the present epoch the ultimate scientific objects are electrons. Each such scientific object has its special relation to each event in nature. Events as thus related to a definite electron are called the 'field' of that object. The relations of the object to different parts of the field are interconnected; and, when the relationship of the object to certain parts of the field is known, its relationship to the remaining parts can be calculated.

As here defined the field of an electron extends through all time and all space, each event bearing a certain character expressed by its relation to the electron. As in the case of other objects, the electron is an atomic unity, only mediately in space and in time by reason of its specific relations to events. This field is divisible into two parts, namely, the 'occupied' events and the 'unoccupied' events. The occupied event corresponds to the situation of a physical object. In order to express these relations of an electron to events with sufficient simplicity, the method of extensive abstraction [cf. Part III] has to be employed. The success of this method depends on the principle of convergence to simplicity with diminution of extent. The result is to separate off the temporal and spatial properties of events. The relations of electrons to events can be expressed in terms of spatial positions and spatial motions at all instants throughout the whole of time.

25·6 In terms of space and time (as derived by the method of extensive abstraction) the situation of a physical object shrinks into its spatial position at an instant together with its associated motion. Also an event occupied by an electron shrinks into the position at an instant of the electric charge forming its nucleus, together with its associated motion. But the quantitative charge is entirely devoid of character apart from its associated field; it expresses the character of the occupied events which is due to their relation to the electron. Its permanence of quantity reflects the permanence which is recognised in the electron, considered for itself alone.

25·7 The 'unoccupied' events possess a definite character expressive of the reign of law in the creative advance of nature, i.e. in the passage of events. This type of character of events unoccupied by the electron is also shared by the occupied events. It expresses the rôle of the electron as an agency in the passage of events. In fact the electron is nothing else than the expression of certain permanent recognisable features in this creative advance.

Thus the character of event *e* which it receives from electron *A*, which does not occupy it, is one of the influences which govern the change of electron *B*, which does occupy *e*, into the occupation of other events succeeding *e*. The complete rule of change for *B* can be expressed in terms of the complete character which *e* receives from its relations to all the electrons in the universe.

25·8 The connectedness of the characters which events receive from a given electron is expressed by the notion of transmission, namely the characters are transmitted from the occupied events according to a regular rule, which depends on the continuity of events arising from their mutual relations of extension. This transmission through events is expressible as a transmission through space with finite velocity.

25·9 Thus in an event unoccupied by it an electron is discerned only as an agent modifying the character of that event; whereas in an event occupied by it the electron is discerned as itself acted on, namely the character of that event governs the fate of the electron. Thus in a sense there is no action at a distance; for the fate of each electron is wholly determined by the event it occupies. But in a sense there is action at a distance, since the character of any event is modified (to however slight a degree) by any other electron, however separated by intervening events. This action at a distance is in its turn limited to being a transmission through the intervening events.

26. Duality of Nature. 26·1 There are two sides to nature, as it were, antagonistic the one to the other, and yet each essential. The one side is development in creative advance, the essential becomingness of nature. The other side is the permanence of things, the fact that nature can be recognised. Thus nature is always a newness relating objects which are neither new nor old.

26·2 Perception fades unless it is equally stimulated from both sides of nature. It is essentially apprehension of the becomingness of nature. It requires transition, contrast, and newness, and immediacy of happening. Thus essentially perception is an awareness of events in the act of passing into what has never yet been. But equally perception requires recognition. Now electrons —in so far as they are ultimate scientific objects and if they are such objects—do not satisfy the complete condition for recognisability.

26·3 Such ultimate scientific objects embody what is ultimately permanent in nature. Thus they are the objects whose relations in events are the unanalysable expression of the order of nature. But the recognition in perception requires the recurrence of the ways in which events pass. This involves the rhythmic repetition of the characters of events. This permanence of rhythmic repetition is the essential character of molecules, which are complex scientific objects. There is no such thing as a molecule at an instant. A molecule requires a minimum of duration in which to display its character. Similarly physical objects are steady complexes of molecules with an average permanence of character throughout certain minimum durations.

26·4 Thus the recognition which is involved in perception is the reason for the importance in physical science of Lorentz's hierarchy of microscopic and macroscopic equations.

26·5 The further consideration of objects, in particular their instantaneous spatial positions and the quantitative distribution of material through space, is resumed in Part IV, after the theory of space and time has been established.

PART III

THE METHOD OF EXTENSIVE ABSTRACTION

CHAPTER VIII

PRINCIPLES OF THE METHOD OF EXTENSIVE ABSTRACTION

27. The Relation of Extension, Fundamental Properties.
27·1 The fact that event a extends over event b will be expressed by the abbreviation aKb. Thus 'K' is to be read 'extends over' and is the symbol for the fundamental relation of extension.

27·2 Some properties of K essential for the method of extensive abstraction are,

(i) aKb implies that a is distinct from b, namely, 'part' here means 'proper part':

(ii) Every event extends over other events and is itself part of other events: the set of events which an event e extends over is called the set of parts of e:

(iii) If the parts of b are also parts of a and a and b are distinct, then aKb:

(iv) The relation K is transitive, i.e. if aKb and bKc, then aKc:

(v) If aKc, there are events such as b where aKb and bKc:

(vi) If a and b are any two events, there are events such as e where eKa and eKb.

It follows from (i) and (iv) that aKb and bKa are inconsistent. Properties (ii) and (v) and (vi) together postulate something like the existence of an ether; but it is not necessary here to pursue the analogy.

28. Intersection, Separation and Dissection. 28·1 Two events 'intersect' when they have parts in common. Intersection, as thus defined, includes the case when one event extends over the other, since K is transitive. If every intersector of b also intersects a, then either aKb or a and b are identical.

Events which do not intersect are said to be 'separated.' A 'separated set' of events is a set of events of which any two are separated from each other.

28·2 A 'dissection' of an event is a separated set such that the set of intersectors of its members is identical with the set of intersectors of the event. Thus a dissection is a non-overlapping exhaustive analysis of an event into a set of parts, and conversely the dissected event is the one and only event of which that set is a dissection. There will always be an indefinite number of dissections of any given event.

If aKb, there are dissections of a of which b is a member. It follows that if b is part of a, there are always events separated from b which are also parts of a.

29. The Junction of Events. 29·1 Two events x and y are 'joined' when there is a third event z such that (i) z intersects both x and y, and (ii) there is a dissection of z of which each member is a part of x, or of y, or of both.

The concept of the continuity of nature arises entirely from this relation of the junction between two events. Two joined events are continuous one with the other. Intersecting events are necessarily joined; but the notion of junction is wider than that of intersection, for it is possible for two separated events to be joined.

Two events which are joined have that relation to each other necessary for the existence of one event which extends over them and over no extraneous events. Two events which are both separated and joined are said to be 'adjoined.'

29·2 An event x is said to 'injoin' an event y when (i) x extends over y, and (ii) there is some third event z which is separated from x and adjoined to y.

In this definition a property of the boundary of an event first makes its appearance. The assumption that examples of the relation of injunction hold is a

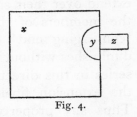

Fig. 4.

long step towards a theory of such boundaries, as the annexed diagram illustrates. It is important to note that injunction has been defined purely in terms of extension.

If xKy and z is separated from x and adjoins y, then z adjoins x.

29·3 Injunction and adjunction are the closest types of boundary union possible respectively for an event with its part and for a pair of separated events. The geometry for events is four-dimensional, but in the three-dimensional analogue such a surface union for a pair of volumes would be the existence of a finite area of surface in common.

[Note that spatial diagrams, such as the one above, are to some extent misleading in that they emphasise the spatial character of events at the expense of their temporal character. The temporal character is very far from being represented by an extra dimension producing an ordinary four-dimensional euclidean geometry.]

30. Abstractive Classes. 30·1 A set of events is called an 'abstractive class' when (i) of any two of its members one extends over the other, and (ii) there is no event which is extended over by every event of the set.

The properties of an abstractive class secure that its members form a series in which the predecessors extend over their successors, and that the extension of the members of the series (as we pass towards the 'converging end' comprising the smaller members) diminishes without limit; so that there is no end to the series in this direction along it and the diminution of the extension finally excludes any assignable event. Thus any property of the individual events which survives throughout members of the series as we pass towards the converging end is a property belonging to an ideal simplicity which is beyond that of any one assignable event. There is no one event which the series marks out, but the series itself is a route of approximation towards an ideal simplicity of 'content.' The systematic use of these abstractive classes is the 'method of extensive abstraction.' All the spatial and temporal concepts can be defined by means of them.

30·2 One class of events—a, say—is said to 'cover' another class of events—β, say—when every member of a extends over some member of β.

If a be an abstractive class and a covers β, then β must have an infinite number of members and there can be no event which is extended over by every member of β. For any member of a, however small, extends over some member of β. The usual case of covering is when both classes, a and β, are abstractive classes; then each member of a, the covering class, extends over the whole converging end of β subsequent to the first member of β which it extends over.

30·3 Two classes of events are called 'K-equal' when each covers the other. Evidently such classes cannot have a finite number of members. K-equality is a relation in which two abstractive classes can stand to each other. The relation is symmetrical and transitive, and every abstractive class is K-equal to itself.

[*Note.* Abstractive classes and the relation of 'covering' can be illustrated by spatial diagrams, with the same caution as to their possibly misleading character.

Consider a series of squares, concentric and similarly situated. Let the lengths of the sides of the successive squares, stated in order of diminishing size, be

$$h_1, h_2, \ldots h_n, \ldots$$

Then each square extends over all the subsequent squares of the set. Also let

$$L_{n \to \infty} h_n = 0;$$

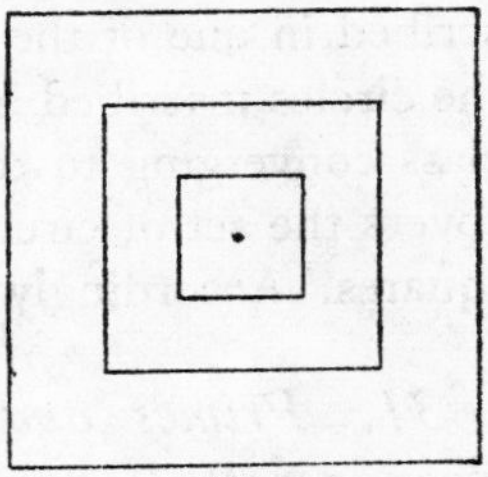

Fig. 5.

namely, let h_n tend to zero as n increases indefinitely. Then the set forms an abstractive class.

Again, consider a series of rectangles, concentric and similarly situated. Let the lengths of the sides of the successive rectangles, stated in order of diminishing size, be $(a, h_1), (a, h_2), \ldots (a, h_n), \ldots$

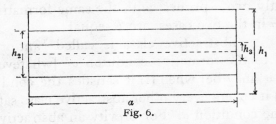

Fig. 6.

Thus one pair of opposite sides is of the same length throughout the whole series. Then each rectangle extends over all the subsequent rectangles. Let h_n tend to zero as n increases indefinitely. Then the set forms an abstractive class.

Evidently the set of squares converges to a point, and the set of rectangles to a straight line. Similarly, using three dimensions and volumes, we can thus diagrammatically find abstractive classes which converge to areas. If we suppose the centre of the set of squares to be the same as that of the set of rectangles, and place the squares so that their sides are parallel to the sides of the rectangles, then the set of rectangles covers the set of squares, but the set of squares does not cover the set of rectangles.

Again, consider a set of concentric circles with their common centre at the centre of the squares, and let each circle be inscribed in one of the squares, and let each square have one of the circles inscribed in it. Then the circles form an abstractive class converging to their common centre. The set of squares covers the set of circles and the set of circles covers the set of squares. Accordingly the two sets are K-equal.]

31. Primes and Antiprimes.

31·1 An abstractive class is called 'prime in respect to the formative condition σ' [whatever condition 'σ' may be] when (i) it satisfies the condition σ, and (ii) it is covered by every other abstractive class satisfying the same condition σ.

For brevity an abstractive class which is prime in respect to a formative condition σ is called a 'σ-prime.' Evidently two σ-primes, with the same formative condition σ in the two cases, are K-equal.

31·2 An abstractive class is called 'antiprime in respect to the formative condition σ' [whatever condition 'σ' may be] when (i) it satisfies the condition σ, and (ii) it covers every other abstractive class satisfying the same condition σ. For brevity an abstractive class which is antiprime in respect to a formative condition σ is called a σ-antiprime. Evidently two σ-antiprimes, with the same formative condition σ in the two cases, are K-equal.

31·3 Let σ be any assigned formative condition, let σ_p be the condition of 'being a σ-prime,' and let σ_a be the condition of 'being a σ-antiprime.' Thus an abstractive class, which satisfies the condition σ_p, (i) satisfies the condition σ, and (ii) is covered by every other abstractive class satisfying the same condition σ.

Hence any two abstractive classes which satisfy the condition σ_p cover each other. Hence every class which satisfies the condition σ_p is covered by every other class which satisfies the same condition σ_p. That is to say, every such class is a σ_p-prime. Analogously, it is a σ_p-antiprime.

Similarly the σ-antiprimes are the σ_a-primes and σ_a-antiprimes.

A formative condition σ will be called 'regular for primes' when (i) there are σ-primes and (ii) the set of abstractive classes K-equal to any one assigned σ-prime is identical with the complete set of σ-primes; and σ will be called 'regular for antiprimes' when (i) there are σ-antiprimes and (ii) the set of abstractive classes K-equal to any one assigned σ-antiprime is identical with the complete set of σ-antiprimes. Thus if σ be a formative condition regular for primes, the set of σ-primes is the same as the set of abstractive classes K-equal to σ-primes; and if σ be a formative condition regular for antiprimes, the set of σ-antiprimes is the same as the set of abstractive classes K-equal to σ-antiprimes.

31·4 Errors arise unless we remember the existence of some exceptional abstractive classes. Since we assume that each event has a definite demarcation we know that the laws of nature ordinarily assumed in science will issue in ascribing to each event a definite boundary which will be a spatial surface prolonged into three dimensions by reason of its time-extension. Thus the possibilities of the spatial contact of surfaces are reproduced in the three-dimensional boundaries of events. Abstractive classes exist whose converging ends converge to elements [instantaneous points, or routes, or etc.] on the surface of one of the members of the class. In such a case, as we pass down the abstractive class towards its converging end, after some definite member x of the class the remaining members, all extended over by x, have some form of internal contact with the boundary of x. The closest form of such contact is to be injoined in x. But there will also be more abstract types of point-contact or of line-contact which we have not defined here, but know about from their occurrence in geometry. If we merely exclude such cases without explicit definition, we are really appealing to fundamental relations and properties which have not been explicitly recognised. We must use definitions based solely upon those properties of the relation K which have been made explicit. We cannot explicitly take account of point-contact till points have been defined.

32. Abstractive Elements. 32·1 A 'finite abstractive element deduced from the formative condition σ' is the set of events which are members of σ-primes, where σ is a formative condition regular for primes. The element is said to be 'deduced' from its formative condition σ.

An 'infinite abstractive element deduced from the

formative condition σ' is the set of events which are members of σ-antiprimes, where σ is a formative condition regular for antiprimes. The element is said to be 'deduced' from its formative condition σ.

The abstractive elements are the set of finite and infinite abstractive elements.

32·2 An abstractive element deduced from a regular formative condition σ is such that every abstractive class formed out of its members either covers all σ-primes [element finite] or is covered by all σ-antiprimes [element infinite]. Thus it represents a set of equivalent routes of approximation guided by the condition that each route is to satisfy the condition σ.

32·3 An abstractive element will be said to 'inhere' in any event which is a member of it. Two elements such that there are abstractive classes covered by both are said to 'intersect' in those abstractive classes.

One abstractive element may cover another abstractive element. The elements of the utmost simplicity will be those which cover no other abstractive elements. These are elements which in euclidean phrase may be said to be 'without parts and without magnitude.' It will be our business to classify some of the more important types of elements. The elements of the greatest complexity will be those which can cover elements of all types. These will be 'moments.'

A point of nomenclature is important. We shall name individual abstractive elements by capital latin letters, classes of elements by capital or small latin letters, and also, as heretofore, events by small latin letters. K will continue to denote the fundamental relation of extension from which all the relations here considered are derived.

CHAPTER IX

DURATIONS, MOMENTS AND TIME-SYSTEMS

33. Antiprimes, Durations and Moments. 33·1 Among the constants of externality discussed in Part II was the reference of events to durations which are, in a sense, complete wholes of nature. A duration has thus in some sense an unlimited extension, though it is bounded in its temporal extent. Although we have not yet in our investigation of K distinguished between spatial and temporal extension, durations can nevertheless be defined in terms of K by this unlimited aspect of their extents. Namely, we assume that there are no other events with the same unlimited property. Accordingly, any abstractive class which is composed purely of durations can only be covered by abstractive classes which also are composed purely of durations.

33·2 An abstractive class a is called an 'absolute antiprime' when a is itself one of the antiprimes which satisfy the formative condition of covering a. In other words, an absolute antiprime is an abstractive class which covers every abstractive class which covers it.

If an abstractive class be an absolute antiprime, it is evident that the formative condition of 'covering it' is regular for antiprimes. Thus the set of events which are members of the absolute antiprimes which cover some one assigned absolute antiprime constitutes an abstractive element. Such an element will be called a 'moment.' Thus a moment is an abstractive element deduced from the condition of covering an absolute antiprime.

Only events of a certain type can be members of an absolute antiprime, namely events which in Part II have been called 'durations.' Only durations can extend over durations, and accordingly all the members of a moment are durations.

33·3 We may conceive of a duration as a sort of temporal thickness (or, slab) of nature*. In an absolute

* The slab of nature forming a duration is limited in its temporal dimension and unlimited in its spatial dimensions. Thus it represents a finite time and infinite space. For example let the horizontal

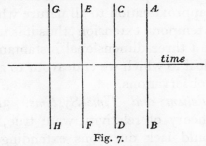

Fig. 7.

line represent the time; and assume nature to be spatially one-dimensional, so that an unlimited vertical line in the diagram represents

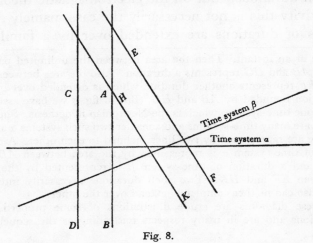

Fig. 8.

antiprime we have a series of temporal thicknesses successively packed one inside the other and converging towards the ideal of no thickness. An absolute antiprime indicates the ideal of an extensionless moment of time.

The set of moments which inhere in a duration are completely characteristic of that duration, and vice versa. A moment is to be conceived as an abstract of all nature at an instant. No abstractive element can cover a moment except that moment itself. A moment is a route of approximation to all nature which has lost its (essential) temporal extension; thus it is nature under the aspect of a three-dimensional instantaneous space. This is the ideal to which we endeavour to approximate in our exact observations.

34. Parallelism and Time-Systems. 34·1 If the Newtonian theory of relativity were true, no pair of durations would lack durations extending over both of them, namely larger durations including both the given durations. But on the electromagnetic theory of relativity this is not necessarily the case, namely some pairs of durations are extended over by a family of

---

space at an instant. Then the area between the unlimited parallel lines *AB* and *HG* represents a duration. Also the area between *CD* and *EF* represents another duration which is extended over by the duration bounded by *AB* and *HG*. But in fig. 7 we have assumed only one time-system, which is the Newtonian hypothesis. Suppose there are many time-systems and consider two such systems α and β. These are represented by two lines inclined to each other. A duration of time-system α is represented by the area between *AB* and *CD*, and a duration of time-system β is represented by the area between *EF* and *HK*. Two such durations necessarily intersect and also can neither completely extend over the other.

These diagrams are crude illustrations of some properties of durations and are in many respects misleading as the sequel will show.

durations and some are not. We shall adopt the electro-magnetic theory of relativity.

A pair of durations both of which are parts of the same duration are called 'parallel'; and also a pair of moments such that there are durations in which both inhere are called 'parallel.'

Parallelism has the usual properties of transitiveness, symmetry and reflexiveness. Also two durations which do not intersect are parallel; and parallel moments which are not identical never intersect. If two parallel durations intersect there is a duration which is their complete intersection, but there are no durations among the common parts of two durations which are not parallel. Two moments which are not parallel necessarily intersect.

34·2 Two durations which are parallel to the same duration are parallel to each other; thus it is evident that each absolute antiprime and each moment must be composed of parallel durations.

A 'family of parallel durations' is formed by all the durations parallel to a given duration, including that duration itself. Evidently any two members of such a family are parallel, and no duration out of the family is parallel to any duration of the family.

Analogously to such families of parallel durations, there are families of parallel moments, with the property that no two moments of the same family intersect and that any moment out of a given family intersects every moment belonging to the family.

The durations which are the members of the various moments of a given family of moments themselves form a family of parallel durations. Thus corresponding to a family of parallel durations there is one and only one family of parallel moments; and corresponding to a family of parallel moments there is one and only one family of parallel durations. A pair of such corresponding families, one of durations and the other of moments, form the 'time-system' associated with either of the two families.

Evidently each duration belongs to one and only one family of parallel durations; and thus each duration belongs to one and only one time-system. Also each moment belongs to one and only one family of parallel moments; and thus each moment belongs to one and only one time-system. Thus two distinct time-systems have no durations in common and no moments in common. But every event not a duration is contained in some durations of any given time-system. Furthermore there will be a minimum duration in a given time-system which is the duration 'when' the event happened in that time-system; namely, the minimum duration has the properties (i) that it extends over the event and (ii) that every duration which is part of it intersects the event.

34·3   The moments of a time-system are arranged in serial order in this way:

(i) A duration belonging to a time-system is 'bounded' by a moment of the same time-system when each duration in which that moment inheres intersects the given duration and also intersects events separated from the given duration:

(ii) Every duration has two such bounding moments, and every pair of parallel moments bound one duration of that time-system:

(iii) A moment B of a time-system 'lies between' two moments A and C of the same time-system when B inheres in the duration which A and C bound:

(iv) This relation of 'lying between' has the following properties which generate continuous serial order in each time-system, namely,

(α) Of any three moments of the same time-system, one of them lies between the other two:

(β) If the moment B lies between the moments A and C, and the moment C lies between the moments B and D, then B lies between A and D:

(γ) There are not four moments in the same time-system such that one of them lies between each pair of the remaining three:

(δ) The serial-order among moments of the same time-system has the Cantor-Dedekind type of continuity.

Nothing has yet been said about the measurement of the lapse of time. This topic will be considered as part of the general theory of congruence.

35. Levels, Rects, and Puncts. 35·1 The electromagnetic theory of relativity is obviously the more general of the two. It has also the merit of providing definitions of flatness, of straightness, of punctual position, of parallelism, of time-order and spatial order as interconnected phenomena, and (with the help of cogredience) of perpendicularity and of congruence. The theory of extension has also provided the definition of a duration. It is a remarkable fact that the characteristic concepts of time and of geometry should thus be exhibited as arising out of the nature of things as expressed by the two fundamental relations of extension and cogredience. It has already been explained that a moment is the route of approximation towards an

instantaneous three-dimensional whole of nature. The
set of abstractive elements and abstractive classes
covered by both of two non-parallel moments is the
locus which is their common intersection. Such a locus
will be called a 'level' in either moment. A level is in
fact an instantaneous plane in the instantaneous space
of any moment in which it lies. But we reserve the
conventional spatial terms, such as 'plane,' for the
time-less spaces to be defined later. Accordingly the
word 'level' is used here.

35·2 An indefinite number of non-parallel moments
will intersect each other in the same level, forming their
complete intersection; and one level will never be
merely a (logical) part of another level. Let three
mutually intersecting moments (M_1, M_2 and M_3, say)
intersect in the levels l_{23}, l_{31}, l_{12}. Then three cases can
arise: *either* (i) the levels are all identical [this will
happen if any two are identical], *or* (ii) no pair of the
levels intersect, *or* (iii) a pair of the levels, say l_{31} and
l_{12}, intersect. In case (i) the three moments are called
'co-level.' In case (ii) there are special relations of
parallelism of levels, to be considered later. In case
(iii) the locus of abstractive elements and abstractive
classes which forms the intersection of l_{31} and l_{12} will
be called a 'rect'; let this rect be named r_{123}. Then
r_{123} is also the complete intersection of l_{12} and l_{23}, and
of l_{23} and l_{31}, and of the three moments M_1, M_2, M_3.
When three moments have a rect as their complete
intersection they are called 'co-rect.' A rect is an instantaneous straight line in the instantaneous three-
dimensional space of any moment in which it lies.
But, as before, the conventional space-nomenclature
is avoided in connection with instantaneous spaces.

35·3 For four distinct moments there are four possible cases in respect to their intersection. In case (i) there is no common intersection: in case (ii) there is a common intersection which is a level: in case (iii) there is a common intersection which is a rect: in case (iv) there is a common intersection which is neither a rect nor a level; in this case the common intersection will be called a 'punct.'

Consider four moments M_1, M_2, M_3, M_4 which constitute an instance of case (iv). Let l_{12} be the level which is the intersection of M_1 and M_2, and let r_{234} be the rect which is the intersection of M_2, M_3, M_4. Then the rect r_{234} does not lie in the level l_{12}. The rect r_{234} intersects the level l_{12} in the common intersection of the four moments. This common intersection is an instantaneous point in the instantaneous spaces of the moments. In accordance with our practice of avoiding the conventional spatial terms when speaking of an instantaneous space, we have called this intersection a 'punct.' Since space is three-dimensional, any moment either covers every member of a given punct or covers none of its members. A punct represents the ideal of the maximum simplicity of absolute position in the instantaneous space of a moment in which it lies.

35·4 It is tempting, on the mathematical analogy of four-dimensional space, to assert the existence of unlimited events which may be called the complete intersections of pairs of non-parallel durations. It is dangerous however blindly to follow spatial analogies; and I can find no evidence for such unlimited events, forming the complete intersections of pairs of intersecting durations, except in the excluded case of parallelism when the complete intersection (if it exist) is itself a duration. Accordingly, apart from parallelism, it may be assumed that the events extended over by a pair of intersecting durations are all finite events. No change in the sequel is required if the existence of such infinite events be asserted.

36. Parallelism and Order. 36·1 Two levels which are the intersections of one moment with two parallel moments are called 'parallel.' Two parallel levels do not intersect, and conversely two levels in the same moment which do not intersect are parallel.

In any moment there will be a complete system of levels parallel to a given level in that moment, and such levels will be parallel to each other.

Similarly 'parallel' rects are defined by the intersection of parallel levels with a given level, all in one moment. Thus within any moment the whole theory of euclidean parallelism (so far as it is non-metrical) follows, and need not be further elaborated except to note the existence of parallelograms.

36·2 The definitions of parallel levels and of parallel rects can be extended to include levels and rects which are not co-momental:

(i) Two levels, l and l', are parallel if l is the intersection of moments M_1 and M_2, and l' of moments M_1' and M_2', where M_1 is parallel to M_1' and M_2 to M_2':

(ii) Two rects, r and r', are parallel if r is the intersection of co-momental levels l_1 and l_2, and r' of co-momental levels l_1' and l_2', where l_1 is parallel to l_1' and l_2 to l_2'.

A moment and a rect which do not intersect are parallel. A rect either intersects a moment in one punct, or is parallel to it, or is contained in it.

36·3 The essential characteristic of space is bound up in what may be termed the 'repetition property' of parallelism. This repetition property is an essential element in congruence as will be seen later; also the homogeneity of space depends on it. Examples of the repetition property are as follows: if a rect intersects any moment in one and only one punct, then it intersects each moment of that time-system in one and only one punct: if a level intersects any moment in one and only one rect, then it intersects any moment of that time-system in one and only one rect. But we must not apply the theory of repetition in parallelism mechanically without attention to the nature of the property concerned. For example, if a rect is incident in a moment, it does not intersect any other moment of the same time-system, and therefore *à fortiori* is not incident in any of them; and analogously for a level incident in a moment.

36·4 Puncts on a rect have an order which is derivative from the order of moments in a time-system and which connects the orders of various time-systems. The puncts on any given rect r will respectively be incident in the moments of any time-system α to which the rect is not parallel. Any moment of α will contain one punct of r, and any punct of r will lie in one moment of α. Thus the puncts of r have derivatively the order of the moments of α. Again let β be another such time-system. Then the puncts of r have derivatively the order of the moments of β. But it is found that these two orders for puncts on r are identical, namely there is only one order for the puncts on r to be obtained in this way. By means of these puncts on rects the orders of moments of different time-systems are correlated.

Thus the existence of order in the instantaneous spaces of moments is explained; but the theory of congruence has not yet been entered upon.

36·5 The set of puncts, rects and levels in any one moment thus form a complete three-dimensional euclidean geometry, of which the meaning of the me ri-cal properties has not yet been investigated. It is no necessary here to enunciate the fundamental propositions [such as two puncts defining a rect, and so on] from which the whole theory can be deduced so far as metrical relations are not concerned.

CHAPTER X

FINITE ABSTRACTIVE ELEMENTS

37. Absolute Primes and Event-Particles. 37·1 It follows from the principles of convergence to simplicity with diminution of extent that, for exhibiting the relations between events in their utmost simplicity, abstractive elements of minimum complexity are required, that is, elements which converge towards the ideal of an atomic event. This requisite exacts that the formative condition from which the 'atomic' element is deduced should be such as to impose the minimum of restriction on convergence.

37·2 An abstractive class which is prime in respect to the formative condition of 'covering all the elements and abstractive classes constituting some assigned punct' is called an 'absolute prime.'

Evidently the condition satisfied by an absolute prime is regular for primes. The abstractive element deduced from an absolute prime is called an 'event-particle.' An event-particle is the route of approximation to an atomic event, which is an ideal satisfied by no actual event.

An abstractive class which is antiprime in respect to the formative condition of 'being a member of some assigned punct' is evidently an absolute prime. In fact this set of antiprimes is identical with the set of abso ute primes.

An event-particle is an instantaneous point viewed in the guise of an atomic event. The punct which an event-particle covers gives it an absolute position in the instantaneous space of any moment in which it lies. Event-particles on a rect lie in the order derived from the puncts which they cover.

37·3 The complete set of event-particles inhering in an event will be called the set 'analysing' that event. A set of event-particles can only analyse one event, and an event can be analysed by only one set of event-particles.

An event-particle 'bounds' an event x when every event in which the event-particle inheres intersects both x and events separated from x. The set of event-particles bounding an event is called the 'boundary' of that event. A boundary can only bound one event and every event has a boundary.

Event-particles which neither inhere in an event nor bound it are said to lie 'outside' it.

The existence of boundaries enables the contact of events to be defined, namely, events are in 'contact' when their boundaries have one or more event-particles in common. The adjunction of events implies contact but not vice versa; since adjunction requires that a solid of the boundaries should be in common. But we define the notion of solid by means of that of adjunction, and not conversely.

37·4 If A and B are distinct event-particles, there are events separated from each other in which A and B respectively inhere.

Two events intersect if there are event-particles each inhering in both events; and conversely, there are event-particles inhering in both events if they intersect.

37·5 The fact that the instantaneous geometry within a moment is three-dimensional leads to the conclusion that the geometry for all event-particles will be four-dimensional. It is to be noted however that the straight lines for this four-dimensional geometry have so far only been defined for event-particles which are co-momental, namely the rects. Event-particles which are not co-momental will be called 'sequent.' Straight lines of the four-dimensional space joining sequent event-particles will be defined in Chapter XI.

37·6 The theory of contact is based on the four-dimensionality of the geometry of event-particles. Some results of that datum are now to be noted.

A 'simple' abstractive class is an abstractive class for which there is no one event-particle on the boundaries of all those members of the converging end, which succeed some given member of the class; namely, for a simple abstractive class there is no one event-particle at which all members of the converging end have contact.

Absolute antiprimes and absolute primes are simple abstractive classes. The 'atomic' property of an absolute prime is expressed by the theorem, that an absolute prime is a simple abstractive class which is covered by every simple abstractive class which it covers. The property of 'instantaneous completeness' exhibited by an absolute antiprime is expressed by the theorem, that an absolute antiprime is an abstractive class which covers every abstractive class that covers it.

38. *Routes.* 38·1 Event-particles are abstractive elements of atomic simplicity. Routes are abstractive elements in which is found the first advance towards increasing complexity.

A 'linear' abstractive class is a simple abstractive class (λ) which (i) covers two event-particles ρ_1 and ρ_2

(called the end-points), and (ii) is such that no selection of the event-particles which it covers can be the complete set of event-particles covered by another simple abstractive class, provided that the selection comprises ρ_1 and ρ_2 and does not comprise all the event-particles covered by λ. The condition (i) secures that a linear abstractive class converges to an element of higher complexity than an event-particle; and the condition (ii) secures that it has the linear type of continuity.

A 'linear prime' is an abstractive class which is prime in respect to the formative condition of (i) being covered by an assigned linear abstractive class covering two assigned end-points and (ii) being itself a linear abstractive class covering the same assigned end-points. This formative condition is evidently regular for primes.

A 'route' is the abstractive element deduced from a linear prime. The two assigned event-particles which occur as end-points in the definition of the linear prime from which a route is deduced are called the 'end-points' of that route. A route is said to lie between its end-points.

38·2 A route is a linear segment, straight or curved, between two event-particles, co-momental or sequent. There are an indefinite number of routes between a given pair of event-particles as end-points. A route will cover an infinite number of event-particles in addition to its end-points. The continuity of events issues in a theory of the continuity of routes.

If A and B be any two event-particles covered by a route R, there is one and only one route with A and B as end-points which is covered by R.

If A, B and C be any three event-particles covered by a route R, then B is said to 'lie between' A and C

on the route R if B is covered by that route with A and C as end-points which is covered by R.

The particles on any route are arranged in a continuous serial order by this relation of 'lying between' holding for triads of points on it. The necessary and sufficient conditions which the relation must satisfy to produce this serial order are detailed in (iv) of 34·3.

38·3 A route may, or may not, be covered by a moment. If it is so covered it is called a 'co-momental route.' A 'rectilinear route' is a route such that all the event-particles which it covers lie on a rect. In a rectilinear route the order of the event-particles on the rect agrees with the order of the event-particles as defined by the relation of 'lying between' as defined for the route.

Between any two event-particles on a rect there is one and only one rectilinear route. If A and B be two event-particles on a rect, the rectilinear route between them can also be defined as the element deduced from the prime with the formative condition of being a simple abstractive class which covers A and B and all the event-particles between A and B on the rect.

38·4 Among the routes which are not co-momental, the important type is that here named 'kinematic routes.' A 'kinematic route' is a route (i) whose end-points are sequent and (ii) such that each moment, which in any time-system lies between the two moments covering the end-points, covers one and only one event-particle on the route, and (iii) all the event-particles of the route are so covered.

The event-particles covered by a kinematic route represent a possible path for a 'material particle.' But this anticipates later developments of the subject, since the concept of a 'material particle' has not yet been defined.

39. Solids. **39·1** A 'solid prime' is a prime with the formative condition of being a simple abstractive class which covers all the event-particles shared in common by both boundaries of two adjoined events. This formative condition is evidently regular for primes. A 'solid' is the abstractive element deduced from a solid prime.

39·2 If two event-particles are covered by a solid, there are an indefinite number of routes between them covered by the same solid.

A solid may or may not be covered by a moment. If it is so covered, it is called 'co-momental.'

A solid which is not co-momental is called ' vagrant.' The properties of vagrant solids are assuming importance in connection with Einstein's theory of gravitation; the consideration of these properties is not undertaken in this enquiry. Co-momental solids are also called 'volumes.' Volumes are capable of a simpler definition which is given in the next article.

40. Volumes. **40·1** A 'volume prime' is a prime with the formative condition of being a simple abstractive class which covers all the event-particles inhering in an assigned event and covered by an assigned moment. If there are no such particles, there will be no corresponding volume prime. This formative condition is evidently regular for primes.

A 'volume' is the abstractive element deduced from a volume prime. A volume is thus the section of an event made by a moment.

40·2 Any volume is covered by the assigned moment of which mention occurs in its definition. Thus every volume (as here defined) is co-momental. Also a volume only covers those event-particles of which mention occurs in its definition. This set of event-particles is completely characteristic of the volume, and may be considered as the volume conceived as a locus of event-particles.

In exactly the same way a solid, or a route, is completely defined by the event-particles which it covers and vice versa. Thus solids and routes can be conceived as loci of event-particles.

The concrete event itself is also defined by (or, analysed by) the event-particles inhering in it, and such a set of event-particles defines only one event. Thus an event can be looked on as a locus of event-particles. An event e' which is part of an event e is defined in this way by a set of event-particles which are some of the set defining e. This fact is the reason for the confusion of the logical 'all' and 'some' with the physical 'whole' and 'part' which apply solely to events. An event is also uniquely defined by the set of event-particles which form its boundary.

CHAPTER XI

POINTS AND STRAIGHT LINES

41. Stations. 41·1 The fact that an event is 'co-gredient' with a duration is a fundamental fact not to be explained purely in terms of extension. It has been pointed out in Part II that the exact concept of cogredience is 'Here throughout the duration' or 'There throughout the duration.' Let this fundamental relation of finite events to durations be denoted by 'G,' and let 'aGb' mean 'a is a finite event which is cogredient with the duration b.'

41·2 A 'stationary prime' within a duration b is a prime whose formative condition (σ) is that of being a simple abstractive class, such that each of its members extends over events which (i) are inhered in by some assigned event-particle P inherent in b and (ii) have the relation G to b. This formative condition is regular for primes. A 'station' within a duration b is the abstractive element deduced from a stationary prime within b.

41·3 Each event-particle in a duration is covered by one and only one station in that duration; and any event-particle covered by a station can be taken as the 'assigned event-particle' of the formative condition, inherent in every event which is a member of the station. Every station is a route; and also every station in a duration intersects every moment of that duration [i.e. inherent in it] in one and only one event-particle, and intersects no other moments of that time-system. It will be noted that a station is associated with a definite time-system, namely the time-system corresponding to its duration.

41·4 A station of one time-system either does not intersect a station of another time-system or intersects it in one event-particle only. Thus stations belong to the type of routes which have been denominated 'kinematic routes.' Each station exhibits an unchanging meaning of 'here' throughout the duration in which it is a station; namely, every event-particle in a station is 'here' in the duration in the same sense of 'here' as for every other event-particle in that station.

42. Point-Tracks and Points. 42·1 Consider all the durations belonging to one time-system. Of these durations some intersect each other, and some are parts of others. Thus any event-particle P is covered by many durations of this time-system, and lies in stations corresponding to these durations. We have now to consider the relations to each other of these various stations, each containing P. The fundamental theorem is as follows: If d and d' be durations of the same time-system, and d extends over d', and if P be an event-particle inhering in d', and s and s' be the stations of P in d and d' respectively, then s covers s'. In other words used in less technical senses, If d' be part of d, then s' is part of s.

42·2 Any given station s in a duration d can thus be indefinitely prolonged throughout the time-system to which d belongs. For let d_1 be any other duration of the same time-system which intersects d in the duration d' and also extends beyond d. Then the part of s which is included in d', namely s' (say), is a station in d'. Also there is one and only one station in d_1, s_1 (say), which covers s'; and no other station in d_1 covers any

event-particle of s'. In this way the station s is prolonged in the time-system by the addition of the station
s_1, and so on indefinitely. The complete locus of event-
particles thus defined by the indefinite prolongation
of a station throughout its associated time-system is
called a 'point-track.'

A point-track intersects any moment of any time-
system in one and only one event-particle.

42·3 Each point-track has a unique association with
the time-system in which the routes lying on it are
stations. A point-track is called a 'point' in the 'space
of its associated time-system.' This space of a time-
system is called 'time-less' because its points have no
special relation to any one moment of its associated
time-system.

Each event-particle is contained in one and only one
point of each time-system, and will be said to 'occupy'
such a point. Two points of the same time-system
never intersect; two point-tracks which are respectively
points in the spaces of different time-systems either
do not intersect or intersect in one event-particle only.

Since each point-track intersects any moment in one
and only one event-particle, two co-momental event-
particles cannot lie on the same point-track. A pair
of sequent event-particles lie in one and only one
point-track, apart from exceptional cases when they
lie in 'null-tracks.' Null-tracks are introduced later
in article 45.

42·4 In the four-dimensional geometry of event-
particles it has already been pointed out that rects
have the character of straight lines, but that since
sequent event-particles do not lie on the same rect
there is a missing set of straight lines required to com-

plete the goemetry. Point-tracks [together with the exceptional set of loci termed 'null-tracks'] form this missing set of straight lines for this geometry of event-particles.

The event-particles occupying a point-track have an order derived from the covering moments of any time-system. Those on a null-track have an order derived from routes which it is not necessary to discuss.

43. Parallelism. 43·1 A theory of parallelism holds for point-tracks and can be connected with the analogous theory for rects. Point-tracks which are points in the space of the same time-system are called 'parallel.' Thus a complete family of parallel point-tracks is merely a complete family of points in the space of some time-system. The parallelism of point-tracks is evidently transitive, symmetrical and reflexive. The definition of the parallelism of stations is derived from that of point-tracks.

43·2 The parallelism of point-tracks and the parallelism of rects and moments are interconnected. Let r be any rect in a moment M, and let π be any family of parallel point-tracks. Then a certain set of point-tracks belonging to π will intersect r, and this set will intersect any moment parallel to M in a rect parallel to r. Again let p be any point-track and let ρ be any complete family of parallel rects. Then a certain set of rects belonging to ρ will intersect p; name it ρ_p. Let P be any event-particle on some member of ρ_p; then the point-track containing P and parallel to p will intersect every member of ρ_p.

43·3 A theorem analogous to those of 43·2 also holds for two families of point-tracks. Let p be any point-track and let π be any family of parallel point-tracks to which p does not belong. Then a certain set of point-tracks belonging to π will intersect p; name it π_p. Let P be any event-particle occupying some member of π_p; then the point-track occupied by P and parallel to p will intersect every member of π_p.

This theorem, the theorems of 43·2 and the corresponding theorem for two families of parallel rects are examples of the repetition property of parallelism. It is evident that, given any three event-particles not on one rect or one point-track, a parallelogram can be completed of which the three event-particles are three corners, any one of the event-particles being at the junction of the adjacent sides through the three corners. In such a parallelogram opposite sides are always of the same denomination, namely both rects or both point-tracks; but adjacent sides may be of opposite denominations.

43·4 The event-particles occupying a point p in the time-less space of a time-system α appear at the successive moments of α as successively occupying the same point p. If β be any other time-system, then the point p of the space of α intersects a series of points of the space of β in event-particles which lie on the successive moments of β. These event-particles of p thus occupy a succession of points of β at a succession of moments of β; and we shall find that this locus of points is what is meant by a straight line in the space of β. Thus the point p in the space of α correlates the successive points on a straight line of β with the successive moments of β Thus in the space of β the point p of the space of α appears as exemplifying the kinematical conception of a moving material particle traversing a straight line. It will appear later

POINTS AND STRAIGHT LINES

that, owing to the 'repetition property' of parallelism, the motion is uniform.

44. Matrices. 44·1 A level is obtained by taking a rect *r* and an event-particle *P* co-momental with *r*, and by forming the locus of event-particles on rects through *P* and intersecting *r*, including also particles on the rect through *P* and parallel to *r*.

The same level would be obtained by taking the particles on the rects intersecting *r* and parallel to some one rect through *P* which intersects *r*.

44·2 Analogously to levels, a locus of event-particles called a 'matrix' is obtained by taking a rect *r* and an event-particle *P* which is not co-momental with *r*, and by forming the locus of event-particles on rects or point-tracks through *P* and intersecting *r*, including also the event-particles on the rect through *P* and parallel to *r*.

A 'matrix' is a two-dimensional plane in the four-dimensional geometry of event-particles. Levels and matrices together make up the complete set of such two-dimensional planes, and have the usual properties of such planes which need not be detailed here.

44·3 Matrices are also obtained by taking an event-particle *P* and a point-track *p*, and by forming the locus of event-particles on rects or point-tracks through *P* and intersecting *p*, including also event-particles on the point-track through *P* and parallel to *p*. Any matrix can be generated in either of the two ways. Furthermore matrices can be generated by the use of parallels in the same way as levels are generated as explained in 44·1 and as assumed in 43·4.

45. Null-Tracks. 45·1 The relations between rects and point-tracks are best understood by taking a rect

r and a particle P which is not co-level with r. In this way a matrix is obtained as explained in 44·2.

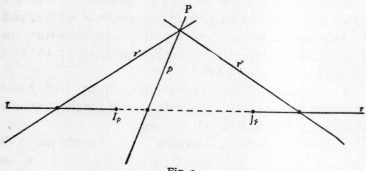

Fig. 9.

Then in respect to P the rect r is divided into three (logical) parts by two event-particles I_p and $\mathcal{J}_p$. The segment between I_p and $\mathcal{J}_p$ has the property that any event-particle on it is joined to P by a point-track [e.g. p in the figure]; and either of the two infinite segments, namely that beyond I_p and that beyond $\mathcal{J}_p$, is such that any event-particle on it is joined to P by a rect [e.g. r' and r'' in the figure]. The above diagram and succeeding diagrams have the defect of representing matrices by levels, and thus of giving the conceptions an undeserved air of paradox.

Again we may take an event-particle P and a point-track p not containing P. In this way a matrix is obtained as explained in 44·3.

Then in respect to P the point-track p is divided by two event-particles I_p and $\mathcal{J}_p$ into three (logical) parts. The segment between I_p and $\mathcal{J}_p$ has the property that any event-particle on it is joined to P by a rect [e.g. r in the figure]; and either of the two infinite segments, respectively beyond I_p and beyond $\mathcal{J}_p$, is such that any event-particle on it is joined to P by a point-track [e.g. p' and p'' in the figure].

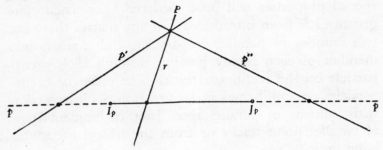

Fig. 10.

45·2 It is evident therefore that a matrix in respect to an event-particle P lying on it is separated into four regions by two loci I_pPI_p' and $\mathcal{J}_pP\mathcal{J}_p'$ which may equally well be termed rects or point-tracks.

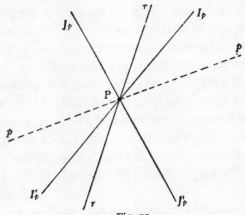

Fig. 11.

The event-particles in the vertically opposed regions $I_pP\mathcal{J}_p$ and $I_p'P\mathcal{J}_p'$ are joined to P by rects; and the event-particles in the vertically opposed regions $I_pP\mathcal{J}_p'$ and $I_p'P\mathcal{J}_p$ are joined to P by point-tracks.

The loci which bound the regions separating point-tracks from rects will be called 'null-tracks.' Their special properties will be considered later when congruence has been introduced. In any matrix there are two families of parallel null-tracks; and there is one member of each family passing through each event-particle on the rectilinear track. The order of event-particles on a null-track is derived from its intersection with systems of parallel rects [not co-momental] or of parallel point-tracks or from the orders on routes lying on it.

46. Straight Lines. 46·1 There is evidently an important theory of parallelism for families of matrices analogous to the theory of parallels for families of levels. The detailed properties need not be elaborated here.

Two matrices may either (i) be parallel, or (ii) intersect in one event-particle only, or (iii) intersect in a rect, or (iv) intersect in a point-track, or (v) intersect in a null-track. For the intersection of two levels only cases (i), (ii) and (iii) can occur; for the intersection of a level and a matrix only cases (ii) and (iii) can occur.

46·2 Each matrix contains various sets of parallel point-tracks. Any one such set is a locus of points in the space of some time-system. Such a locus of points is called a 'straight line' in the space of the time-system.

A matrix which contains the points of a straight line in the space of any time-system α will be called 'an associated matrix for α,' and it is called 'the matrix including' that straight line.

A matrix is an associated matrix for many time-systems, but it is the matrix including only one straight line in each corresponding space. The family of time-systems for which a given matrix is an associated matrix is called a 'collinear' family. A whole family of parallel matrices are associated matrices for the same collinear family of time-systems, if any one matrix of the family is thus associated. In the space of any one time-system the straight lines included by a family of parallel associated matrices are said to be parallel.

46·3 A matrix intersects a moment in a rect. If the moment belong to a time-system with which the matrix is associated, this rect in the moment corresponds to the straight line included by the matrix in the sense that it has one particle occupying each of its points. A rect thus associated with a straight line will be said to 'occupy' it.

Thus the event-particles on a matrix m associated with a time-system a can be exhaustively grouped into mutually exclusive subsets in two distinct ways: (i) They can be grouped into the points of a which lie on m; this locus of points is the included straight line in the space of a, which we will name m_a: (ii) The event-particles on m can be grouped into the sets of parallel rects which are the intersections of m with the moments of a, and thus each of these rects occupies m_a.

46·4 There are three different types of meaning which can be given to the idea of 'space' in connection with external nature. (i) There is the four-dimensional space of which event-particles are the points and the rects and point-tracks and null-tracks are the straight lines. In the geometry of this space there is a lack of uniformity between the congruence theories for rects and for point-tracks, and no such theory for null-tracks. (ii) There are the three-dimensional momentary (instantaneous) spaces in the moments of any time-system α, of which event-particles are the points and rects are the straight lines. The observed space of ordinary perception is an approximation to this exact concept. (iii) There is the time-less three-dimensional space of the time-system α, of which point-tracks are the points and matrices include the straight lines. This is the space of physical science.

There is an exact correlation between the time-less space of a time-system and any momentary space of the same time-system. For any point of the momentary space is an event-particle which occupies one and only one point of the time-less space; and any straight line of the momentary space is a rect which lies in one associated matrix including one straight line of the time-less space, or (in other words) each straight line of the momentary space occupies a straight line of the time-less space.

A time-system corresponds to a consentient set of the Newtonian group, and the time-less space of the time-system is the space of the corresponding consentient group.

CHAPTER XII

NORMALITY AND CONGRUENCE

47. *Normality*. 47·1 A point-track will be said to be 'normal' to the moments of the time-system in the space of which it is a point.

A matrix is said to be 'normal' to the moments which are normal to any of the point-tracks which it contains.

Consider an event-particle P and a matrix m which contains P. Let $\alpha, \beta, \gamma, \ldots$ be the collinear set of time-systems whose points lie in or are parallel to the matrix m. Let P_α, P_β, P_γ, $\ldots$ be the moments of the time-systems α, β, γ, $\ldots$ which contain P. Then the levels $P_{\alpha\beta}$, $P_{\beta\gamma}$, $\ldots$ in which respectively P_α and P_β, P_β and P_γ, etc., intersect are identical, and the event-particle P is the sole event-particle forming the intersection of m and $P_{\alpha\beta}$. Also m intersects each of these moments P_α, and P_β, and P_γ, etc., in rects $r_\alpha, r_\beta, r_\gamma$, etc., respectively. The level $P_{\alpha\beta}$ and the matrix m are said to be mutually 'normal.' It will be noted that any two time-systems, α and β, determine one level and one matrix which are mutually normal and each contain a given event-particle. Corresponding to any level containing P there is one matrix normal to it at P; and corresponding to any matrix containing P there is one level normal to it at P.

If l and m be a level and a matrix normal to each other, then the rects in l will be called normal to the rects and point-tracks in m. A pair of rects which are normal to each other will also be called 'perpendicular' or 'at right-angles.' Two point-tracks can never be normal to each other since no point-track lies on a level. Parallels to normals are themselves normal.

47·2 Continuing the notation of 47·1 we note that the matrix m includes straight lines n_a, n_β, n_γ, etc., of the spaces of a, β, γ, etc., and intersects the moments P_a, P_β, P_γ, etc., in rects r_a, r_β, r_γ, etc., which respectively occupy n_a, n_β, n_γ, etc. The rect r_a contains P and is normal to every rect lying in $P_{a\beta}$. Let r' be any rect containing P and lying in $P_{a\beta}$. Then r' and r_a are mutually normal and both lie in the moment P_a.

The rect r' occupies one straight line in the space of a; name this straight line n'. Then the straight lines n_a and n' will be said to be 'normal' to each other. This definition of the normality of straight lines can be given in general terms thus: Two straight lines in the same space are said to be normal to each other when they are respectively occupied by normal rects lying in the same moment of the corresponding time-system.

47·3 Continuing the notation of 47·2 let l' be the level containing r_a and r'; this level lies in P_a and contains P. Let m' be the matrix normal to l' at P. Then m' intersects $P_{a\beta}$ in a rect r'' which is normal both to r_a and to r'. Thus at an event-particle P in a level $P_{a\beta}$ pairs of mutually normal rects, r' and r'', exist, one of them chosen arbitrarily; and at an event-particle P in a moment P_a triads of mutually normal rects, r_a and r' and r'', exist, with the usual conditions as to freedom of choice.

The correspondence between a momentary space and the time-less space of the same time-system enables us immediately to extend these theorems to pairs of normal straight lines in a plane and to triads of intersecting mutually normal straight lines in three dimensions.

48. Congruence. 48·1 Congruence is founded on the notion of repetition, namely in some sense congruent geometric elements repeat each other. Repetition embodies the principle of uniformity. Now we have found repetition to be a leading characteristic of parallelism; accordingly a close connection may be divined to exist between congruence and parallelism. Furthermore we have just elaborated in outline the principles of normality, pointing out how the property has its origin in the interplay of the relations of extension and co-gredience. But—as we know from experience—a leading property of normality is symmetry, namely, symmetry round the normal. Now symmetry is merely another name for a certain sort of repetition; accordingly congruence and normality should be connected.

We are thus led to look for an expression of the nature of congruence in terms of parallelism and normality, in particular in terms of repetition properties associated with them.

48·2 Congruence, in so far as it is derived from parallelism, is defined by the statements that (i) the opposite sides of parallelograms are congruent to each other, and (ii) routes on the same rect, or on the same point-track, which are congruent to the same route are congruent to each other*.

Also the general law holds that two routes which (as thus defined) are congruent to a third route, are congruent to each other. This law is a substantial theorem as to parallelism, and not a mere consequence of definitions.

* This definition of congruence is given by Profs. E. B. Wilson and G. N. Lewis in their valuable memoir, 'The Space-Time Manifold of Relativity,' *Proc. of the Amer. Acad. of Arts and Sciences*, vol. XLVIII, 1912

But congruence, as thus expressed in terms of parallelism, merely establishes the congruent relation among straight routes on rects belonging to one parallel family, or on point-tracks belonging to one parallel family. For such routes in any one parallel family a system of numerical measurement can be established, of which the details need not be here elaborated. But no principle of comparison has yet been established between the lengths of two routes belonging to different parallel families of rects or belonging to different parallel families of point-tracks. When we can determine equal lengths on any two rects, whether parallel or no, the general principles for space-measurement will have been determined; and when we can determine equal lapses [i.e. lengths] of time on any two point-tracks, whether parallel or no, the general principles for time-measurement will have been determined.

48·3 Congruence as between different parallel families results from the following definition founded on the repetition property [i.e. symmetry] of normality: Let AM and BC be a pair of mutually normal rects intersecting at M, or be a rect and point-track intersecting at M [either AM or BC being the rect] and mutually normal, and let M be the middle event-particle of the straight route BC intervening between the event-particles B and C, then the straight routes AB and AC are congruent to each other.

From the symmetry of normality either both pairs of particles, namely (A, B) and (A, C), are joined by rects, or both pairs are joined by point-tracks, or both pairs by null-tracks. As in the analogous case of congruence derived from parallelism, the transitiveness of congruence expresses a substantial law of nature and not a mere deduction from the terms of the definition.

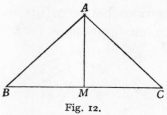

Fig. 12.

48·4 The isosceles triangle of 48·3 must lie either on a level or on a matrix. If it lies on a level, all the straight routes of the figure must lie on rects. But on a matrix a pair of normals cannot be of the same denomination, i.e. not both rects nor both point-tracks. Thus five cases remain over for consideration. These cases are diagrammatically symbolised by the annexed figures where continuous lines represent rects, and dotted lines represent point-tracks.

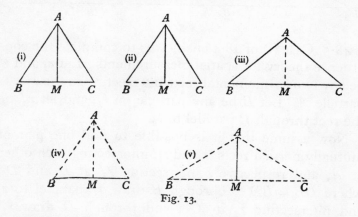

Fig. 13.

Evidently case (i) is the only case in which the triangle lies on a level; the triangles in the remaining four cases lie on matrices.

The relations between the diagrams (ii) and (v) can best be seen by combining them into one figure as in (vi), and the relations between (iii) and (iv) by combining them into one figure as in (vii).

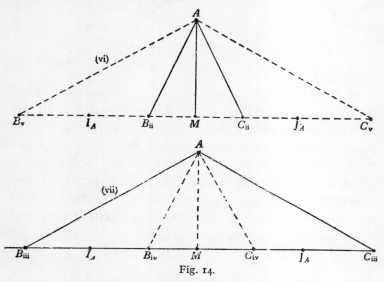

Fig. 14.

48·5 Case (i) of 48·4 enables us to complete the congruence theory for spatial measurements. Let r_1 and r_2 be any two co-momental rects intersecting in the event-particle A. Let B be any particle on r_1, and let r_2' be the rect through B parallel to r_2.

Now assume that it is possible to find one pair of mutually normal rects, r and r', intersecting each other at A, and respectively intersecting r_2' at D and D', where $DB = BD'$. Through B draw r'' parallel to r' and intersecting r_2 in C''; and through D draw r_1' parallel to r_1 and intersecting r_2 in C'.

Then from 48·1, $AC'' = BD'$ and $AC' = BD$. Thus C' and C'' denote the same event-particle. Now

$BM = MC'$. Hence by case (i) of 48·4, $AB = AC'$.
Thus lengths on r_1 and r_2 are comparable. We need
not here consider the theorems, either assumed as independent laws of nature or deduced from previous
assumptions, by which we know that the rectangular
pair (r and r') exist, that C' and C'' coincide and do
not lie on opposite sides of A, and that $BM = MC'$.

48·6   Again if r_1 and r_2 are rects which are not co-
momental and do not lie in parallel moments, their
measurements are still comparable. For two intersecting moments, M_1 and M_2, exist, of which M_1
contains r_1 and M_2 contains r_2. Thus any rect r' in the

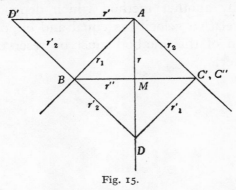

Fig. 15.

level common to M_1 and M_2 has its measurements
comparable both to those on r_1 and to those on r_2; and
thus, by the transitiveness of congruence, the measurements on r_1 and r_2 are comparable. By this procedure
the employment of cases (ii) and (iii) of 48·4 is rendered
unnecessary. Accordingly these cases become theorems
instead of being definitions of congruence as contemplated in their original enunciation. If they had been
taken as definitions, the deduction of 48·5 would still
be possible. But since the figure would now lie in a

matrix, one of r and r' would be a point-track and the other a rect. No very obvious principle then exists by which we could know of the existence of the pair (r and r') such that $DB = BD'$, apart from the assumption of the theorem which we want to prove.

48·7 Cases (iv) and (v) of 48·4 deal with the comparability of time-measurements in different time-systems. The same remarks as those in 48·6 apply; namely that the method of 48·5 could be applied, if independently we could convince ourselves that the requisite pair, r and r' (one a point-track and one a rect), exist.

This comparability of time-measurements will be achieved by another method which depends on the fact that relative velocity is equal and opposite. The explanation of this method must be reserved for the next chapter.

CHAPTER XIII

MOTION

49. *Analytic Geometry.* 49·1 Consider any time-system α: we will term the space of this time-system 'α-space' and its moments 'α-moments'; also the points and straight lines of α-space will be termed 'α-points' and 'α-lines,' and rects and levels which lie in α-moments will be termed 'α-rects' and 'α-levels.' If P be any event-particle, then P_α will denote the α-moment which covers P. If β be any other time-system, there are no β-moments which are also α-moments, and no β-points which are also α-points; but there are α-levels which are also β-levels and α-rects which are also β-rects. For the two moments P_α and P_β intersect in a common level which will be called $P_{\alpha\beta}$. Then rects lying in $P_{\alpha\beta}$ are both α-rects and β-rects. In particular through P in the level $P_{\alpha\beta}$ pairs of mutually normal rects exist, and every rect through P and $P_{\alpha\beta}$ is a member of one such pair.

49·2 Let O be any arbitrarily chosen event-particle, which we will term the origin; and let OO_{at} be the α-point occupied by O; and let OO_{ax}, OO_{ay}, OO_{az} be any triad of mutually rectangular α-rects in the moment O_α, each containing O. In this notation O_{at}, O_{ax}, etc., do not denote any particular entities, but the symbols such as OO_{at} and OO_{ax} are each to be taken as one whole. Let O_{axt} denote the matrix containing OO_{at} and OO_{ax}, with analogous meanings for O_{ayt} and O_{azt}; and let O_{ayz}, O_{azx} and O_{axy} denote respectively the levels containing OO_{ay} and OO_{az}, OO_{az} and OO_{ax},

OO_{ax} and OO_{ay}. Let P be any other event-particle occupying the point OO_{at}, and let PP_{ax}, PP_{ay}, PP_{az} be the a-rects through P respectively parallel to OO_{ax}, OO_{ay}, OO_{az}.

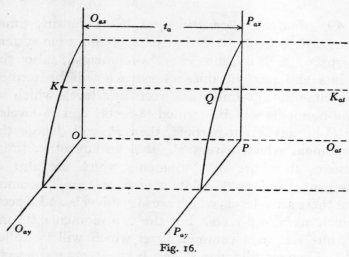

Fig. 16.

In the diagram the third dimension of the moments O_a and P_a, namely the z-dimension, is suppressed, so that these moments are diagrammatically represented as two-dimensional. Point-tracks (in this case a-points) are represented by dotted lines. The diagram has the defect of representing matrices, such as O_{axt}, by levels, and is thus liable to lead to unfounded assumptions.

49·3 Lengths on all rects, whether or no they be a-rects, are measurable in terms of one unit length. But time-lapses between a-moments—or, what is the same thing, time-lapses along a-points—must be measured in a time-unit peculiar to the time-system a, since as yet no means of obtaining congruent time-units in different time-systems has been disclosed. We will suppose at present that in each time-system there is a given arbitrarily chosen unit for time-measurement.

49·4 Let the momentary space of O_a be referred to the three rectangular a-rects OO_{ax}, OO_{ay}, OO_{az} as axes of coordinates; and let the momentary space of P_a be referred to the three rectangular a-rects PP_{ax}, PP_{ay}, PP_{az} as axes of coordinates; and let the time-less space of a [the a-space] be referred to the three rectangular a-lines respectively included in the matrices O_{axt}, O_{ayt}, O_{azt} as axes of coordinates; and let the four-dimensional space of all particles be referred to the four axes consisting of the three a-rects OO_{ax}, OO_{ay}, OO_{az}, and of the a-point OO_{at} as axes of coordinates.

49·5 Let K be any event-particle in the moment O_a, and let K occupy the a-point KK_{at} which intersects the moment P_a in the event-particle Q. Let the lapse of time between the moments O_a and P_a be t_a, where t_a is positive when P_a is subsequent to O_a; and let the coordinates of the a-point KK_{at} in the a-space be (x_a, y_a, z_a). Then the coordinates of K in the momentary space of O_a and of Q in the momentary space of P_a are also (x_a, y_a, z_a). Also the 'a-coordinates' of Q in the four-dimensional space of particles are (x_a, y_a, z_a, t_a); this fact for Q can also be expressed by saying that Q occupies the a-point (x_a, y_a, z_a) at the a-time t_a.

A moment, viewed as a locus of event-particles, is represented by a linear equation in the four coordinates (x_a, y_a, z_a, t_a). But the converse is not true; namely, not every linear equation represents a moment. A pair of linear equations represent a level or a matrix, and three independent linear equations represent a rect or a point-track or a null-track.

49·6 If α and β be any two time-systems, two sets of mutually normal axes, OO_{ax}, OO_{ay}, OO_{az}, OO_{at}, and $OO_{\beta x}$, $OO_{\beta y}$, $OO_{\beta z}$, $OO_{\beta t}$, can be found as in the previous subarticle. But these two sets can evidently be so adjusted that OO_{ay} is identical with $OO_{\beta y}$ and OO_{az} is identical with $OO_{\beta z}$, where the two rects (OO_{ay} and OO_{az}) must both lie in the level $O_{\alpha\beta}$. Then

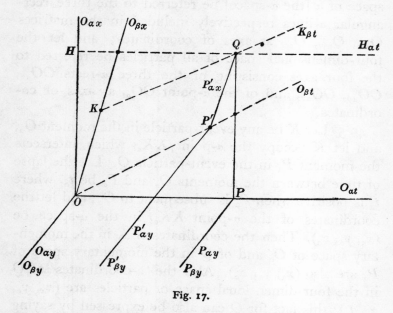

Fig. 17.

the matrix normal to this level at O will be denoted by $O_{\alpha\beta t}$; it contains through O one α-point OO_{at}, one β-point $OO_{\beta t}$, one α-rect OO_{ax}, and one β-rect $OO_{\beta x}$. Then any event-particle is referred to the axes OO_{ax}, OO_{ay}, OO_{az}, OO_{at} for the system α, and to the axes $OO_{\beta x}$, $OO_{\beta y}$, $OO_{\beta z}$, $OO_{\beta t}$ for the system β. Let its α-coordinates be (x_a, y_a, z_a, t_a) and its β-coordinates be $(x_\beta, y_\beta, z_\beta, t_\beta)$, where $y_a = y_\beta$, and $z_a = z_\beta$.

In the diagram, for the sake of simplicity, the particle Q is in the matrix $O_{\alpha\beta t}$; and its coordinates [as in the diagram] in the two systems are $(x_\alpha, \text{o}, \text{o}, t_\alpha)$ and $(x_\beta, \text{o}, \text{o}, t_\beta)$, where PQ (with its proper sign) is x_α, HQ (with its proper sign) is t_α, $P'Q$ (with its proper sign) is x_β, and KQ (with its proper sign) is t_β.

A pair of sets of four axes for α and β allied as described in this subarticle are called 'mutual axes' for the two systems.

49·7 The formulae for transformation from the α-coordinates to the β-coordinates, referred to mutual axes, are obviously of the form

$$x_\beta = \Omega_{\alpha\beta} x + \Omega'_{\alpha\beta} t_\alpha, \; y_\beta = y_\alpha, \; z_\beta = z_\alpha, \; t_\beta = \Omega''_{\alpha\beta} t_\alpha + \Omega'''_{\alpha\beta} x_\alpha \quad \text{(i)},$$

where $\Omega_{\alpha\beta}$, $\Omega'_{\alpha\beta}$, $\Omega''_{\alpha\beta}$, $\Omega'''_{\alpha\beta}$ are constants dependent on the two systems α and β and on the two arbitrarily chosen units of time-lapse in α and β, but evidently not dependent on the arbitrarily chosen set of rectangular rects OO_{ay} and OO_{az} in the level $O_{\alpha\beta}$.

The corresponding (x, t)-equations, interchanging α and β, are

$$x_\alpha = \Omega_{\beta\alpha} x_\beta + \Omega'_{\beta\alpha} t_\beta, \; t_\alpha = \Omega''_{\beta\alpha} t_\beta + \Omega'''_{\beta\alpha} x_\beta \ldots \text{(ii)}.$$

The two pairs of (x, t)-equations, (i) and (ii), must be equivalent. The conditions are

$$\frac{\Omega_{\alpha\beta}}{\Omega''_{\beta\alpha}} = \frac{\Omega''_{\alpha\beta}}{\Omega_{\beta\alpha}} = -\frac{\Omega'_{\alpha\beta}}{\Omega'_{\beta\alpha}} = -\frac{\Omega'''_{\alpha\beta}}{\Omega'''_{\beta\alpha}}$$

$$= \frac{1}{\Omega_{\beta\alpha}\Omega''_{\beta\alpha} - \Omega'_{\beta\alpha}\Omega'''_{\beta\alpha}} = \Omega_{\alpha\beta}\Omega''_{\alpha\beta} - \Omega'_{\alpha\beta}\Omega'''_{\alpha\beta} \ldots \text{(iii)}.$$

Only four out of these five conditions are independent.

50. *The Principle of Kinematic Symmetry.* 50·1 Consider any other time-system π. The π-point (p_π) occupied by $(x_\alpha, y_\alpha, z_\alpha, t_\alpha)$ and the α-point occupied by the same event-particle lie on a matrix m which includes an α-line (m_a) of which every α-point is intersected by p_π. Thus p_π correlates the α-point (x_a, y_a, z_a) with the α-time t_a, and the neighbouring α-point on m_a, namely ($x_a + \dot{x}_a dt_a$, $y_a + \dot{y}_a dt_a$, $z_a + \dot{z}_a dt_a$), with the neighbouring α-time $t_a + dt_a$. In this way π makes every set of α-coordinates of a variable α-point to be a function of t_a; namely it correlates an α-point (x_a, y_a, z_a) with the velocity ($\dot{x}_a$, $\dot{y}_a$, $\dot{z}_a$), which can also be written

$$\left(\frac{dx_a}{dt_a}, \frac{dy_a}{dt_a}, \frac{dz_a}{dt_a} \right).$$

Analogously the same time-system π correlates a β-point (x_β, y_β, z_β) with the velocity ($\dot{x}_\beta$, $\dot{y}_\beta$, $\dot{z}_\beta$), which can be written

$$\left(\frac{dx_\beta}{dt_\beta}, \frac{dy_\beta}{dt_\beta}, \frac{dz_\beta}{dt_\beta} \right).$$

Now the time-system π indicates a definite transference from an event-particle (x_π, y_π, z_π, t_π) to another event-particle (x_π, y_π, z_π, $t_\pi + dt_\pi$) occupying the same π-point (x_π, y_π, z_π), where any mutually normal π-coordinates are employed. The former event-particle is that indicated by (x_a, y_a, z_a, t_a) and by (x_β, y_β, z_β, t_β), and the latter event-particle by ($x_a + \dot{x}_a dt_a$, $y_a + \dot{y}_a dt_a$, $z_a + \dot{z}_a dt_a$, $t_a + dt_a$) and by ($x_\beta + \dot{x}_\beta dt_\beta$, $y_\beta + \dot{y}_\beta dt_\beta$, $z_\beta + \dot{z}_\beta dt_\beta$, $t_\beta + dt_\beta$).

Hence from equations (i) of 49·7

$$\dot{x}_\beta = \frac{\Omega_{a\beta} \dot{x}_a + \Omega'_{a\beta}}{\Omega''_{a\beta} + \Omega'''_{a\beta} \dot{x}_a}, \quad \dot{y}_\beta = \frac{\dot{y}_a}{\Omega''_{a\beta} + \Omega'''_{a\beta} \dot{x}_a}, \quad \dot{z}_\beta = \frac{\dot{z}_a}{\Omega''_{a\beta} + \Omega'''_{a\beta} \dot{x}_a}$$

$$\dots\dots(i).$$

Now π is any time-system. First identify it with α. Then $\dot{x}_a = 0$, $\dot{y}_a = 0$, $\dot{z}_a = 0$. Hence $\dot{y}_\beta = 0$, $\dot{z}_\beta = 0$, and $\dot{x}_\beta$ is the velocity of the time-system α in the space

of β [or, more briefly, the 'velocity of α in β']. Let this velocity be $V_{\beta\alpha}$; it is evidently along the x-axis in the space of β, and
$$V_{\beta\alpha} = \Omega'_{\alpha\beta}/\Omega''_{\alpha\beta} \dots\dots\dots\dots(ii).$$
Again identify the system π with β. Then $\dot{x}_\beta = 0$, $\dot{y}_\beta = 0$, $\dot{z}_\beta = 0$; and hence $\hat{y}_\alpha = 0$, $\dot{z}_\alpha = 0$, and $\dot{x}_\alpha$ is the velocity of β in α. Let this velocity be $V_{\alpha\beta}$; it is along the x-axis in the space of α, and
$$V_{\alpha\beta} = - \Omega'_{\alpha\beta}/\Omega_{\alpha\beta}\dots\dots\dots\dots(iii).$$

50·2 We will now introduce what we will term the 'Principle of Kinematic Symmetry.'

Before enunciating this principle it is necessary to determine a standard method of choosing the positive directions of the axes $OO_{\alpha x}$ and $OO_{\beta x}$ in the matrix $O_{\alpha\beta i}$, and of the axes $OO_{\alpha t}$ and $OO_{\beta t}$. By reference to the figure of subarticle 45·2 it will be seen that, of the four angular regions into which the rects $OO_{\alpha x}$ and $OO_{\beta x}$ divide the matrix $O_{\alpha\beta t}$, two vertically opposite regions include no point-tracks passing through O and the remaining two such regions include point-tracks as well as rects through O. The standard choice of positive directions for $OO_{\alpha x}$ and $OO_{\beta x}$ is such that the two regions bounded one by both positive directions of these axes, and the other by both negative directions, should include only rects passing through O.

The positive directions for $OO_{\alpha t}$ and $OO_{\beta t}$ are settled by the rule that a positive measure of lapse of time should indicate subsequence in the time-order to the moment O_α. This rule is definite because of the ultimate distinction between antecedence and subsequence in time, which has not otherwise been made use of. This standard choice of positive directions along mutual axes for two time-systems will always be adopted.

50·3 The principle of kinematic symmetry has two parts, enunciating consequences which flow from the fact that the time-units in two time-systems α and β are congruent. The first part may be taken as the definition, or necessary and sufficient test, of such congruence.

The first part of the principle can be enunciated as the statement that the measures of relative velocities [i.e. the velocity of β in α and of α in β] are equal and opposite; namely

$$V_{\alpha\beta} + V_{\beta\alpha} = 0 \dots\dots\dots\dots\text{(i)}.$$

The second part is the principle of the symmetry of two time-systems in respect to transverse velocities; namely, if a velocity U in α, normally transverse to the direction of β in α, is represented by the velocity $(V_{\beta\alpha}, U')$ in β, where $V_{\beta\alpha}$ is along the direction of α in β and U' is normally transverse to it, then the same magnitude of velocity U in β, normally transverse to the direction of α in β, is represented by the velocity $(V_{\alpha\beta}, U')$ in α, where $V_{\alpha\beta}$ is along the direction of α in β, and U' is normally transverse to it.

From the first part of the principle, by (ii) and (iii) of 50·1, we deduce

$$\Omega''_{\alpha\beta} = \Omega_{\alpha\beta} \dots\dots\dots\dots\text{(ii)}.$$

In order to apply the second part of the principle we first identify π with $(\dot{x}_\alpha = 0,\ \dot{y}_\alpha = U,\ \dot{z}_\alpha = 0)$, then from (i) and (ii) of 50·1

$$\dot{x}_\beta = V_{\beta\alpha},\ \dot{y}_\beta = U/\Omega''_{\alpha\beta},\ \dot{z}_\beta = 0.$$

Again we identify π with $(\dot{x}_\beta = 0,\ \dot{y}_\beta = U,\ \dot{z}_\beta = 0)$, and by interchanging α and β in the above formulae we find

$$\dot{x}_\alpha = V_{\alpha\beta},\ \dot{y}_\alpha = U/\Omega''_{\beta\alpha},\ \dot{z}_\alpha = 0.$$

Hence by the second part of the principle

$$\Omega''_{\alpha\beta} = \Omega''_{\beta\alpha} \ldots\ldots\ldots\ldots\text{(iii)}.$$

51. Transitivity of Congruence. 51·1 It follows, from (iii) of 49·7, and from (ii) and (iii) of 50·1, and from (i), (ii), (iii) of 50·3, that equations (i) of 50·1 can be written

$$\dot{x}_\beta = \frac{\Omega_{\alpha\beta}(\dot{x}_\alpha - V_{\alpha\beta})}{\Omega_{\alpha\beta} + \Omega'''_{\alpha\beta}\dot{x}_\alpha}, \, \dot{y}_\beta = \frac{\dot{y}_\alpha}{\Omega_{\alpha\beta} + \Omega'''_{\alpha\beta}\dot{x}_\alpha}, \, \dot{z}_\beta = \frac{\dot{z}_\alpha}{\Omega_{\alpha\beta} + \Omega'''_{\alpha\beta}\dot{x}_\alpha}$$
$$\ldots\ldots\text{(i)},$$

where $V_{\alpha\beta} + V_{\beta\alpha} = 0, \, \Omega_{\alpha\beta} = \Omega_{\beta\alpha}, \, \Omega'''_{\alpha\beta} + \Omega'''_{\beta\alpha} = 0,$

$$\Omega_{\alpha\beta}(\Omega_{\alpha\beta} + V_{\alpha\beta}\Omega'''_{\alpha\beta}) = 1 \ldots\ldots\ldots\text{(ii)}.$$

We can now express $\Omega_{\alpha\beta}$ and $\Omega'''_{\alpha\beta}$ in terms of $V_{\alpha\beta}$ and an absolute constant by considering deductions from the transitivity of congruence.

51·2 Let γ be a time-system such that the level $O_{\alpha\gamma}$ contains OO_{ax} and OO_{az}, and let these rects be the axes $OO_{\gamma x}$ and $OO_{\gamma z}$. Then the matrix $O_{\alpha\gamma t}$ contains OO_{ay}, $OO_{\gamma y}$, OO_{at} and $OO_{\gamma t}$. Thus we have obtained a set of mutual axes for α and γ; namely, $(OO_{ax}, OO_{ay}, OO_{az}, OO_{at})$ and $(OO_{\gamma x}, OO_{\gamma y}, OO_{\gamma z}, OO_{\gamma t})$, where OO_{ay} and $OO_{\gamma y}$ now play the part that OO_{ax} and $OO_{\beta x}$ sustain for α and β. Thus the velocities of the time-system π in α and γ are, by (i) of 51·1, connected by

$$\dot{x}_\gamma = \frac{\dot{x}_\alpha}{\Omega_{\alpha\gamma} + \Omega'''_{\alpha\gamma}\dot{y}_\alpha}, \, \dot{y}_\gamma = \frac{\Omega_{\alpha\gamma}(\dot{y}_\alpha - V_{\alpha\gamma})}{\Omega_{\alpha\gamma} + \Omega'''_{\alpha\gamma}\dot{y}_\alpha}, \, \dot{z}_\gamma = \frac{\dot{z}_\alpha}{\Omega_{\alpha\gamma} + \Omega'''_{\alpha\gamma}\dot{y}_\alpha}$$
$$\ldots\ldots\text{(i)}.$$

We have here assumed the congruence of the time-units in α and γ.

Now identify π with γ. Then

$$\dot{x}_\gamma = 0, \, \dot{y}_\gamma = 0, \, \dot{z}_\gamma = 0, \, \dot{x}_\alpha = 0, \, \dot{y}_\alpha = V_{\alpha\gamma}, \, \dot{z}_\alpha = 0.$$

Hence from (i) of $51 \cdot 1$

$$\dot{x}_\beta = -V_{\alpha\beta}, \quad \dot{y}_\beta = V_{\alpha\gamma}/\Omega_{\alpha\beta}, \quad \dot{z}_\beta = 0.$$

But
$$\dot{x}_\beta^2 + \dot{y}_\beta^2 + \dot{z}_\beta^2 = V_{\beta\gamma}^2.$$

Hence
$$V_{\beta\gamma}^2 = V_{\alpha\beta}^2 + V_{\alpha\gamma}^2/\Omega_{\alpha\beta}^2 \ldots \ldots \ldots (\text{ii}).$$

Again identify π with β. Then

$$\dot{x}_\beta = 0, \quad \dot{y}_\beta = 0, \quad \dot{z}_\beta = 0, \quad \dot{x}_\alpha = V_{\alpha\beta}, \quad \dot{y}_\alpha = 0, \quad \dot{z}_\alpha = 0.$$

Hence from (i) of this subarticle

$$\dot{x}_\gamma = V_{\alpha\beta}/\Omega_{\alpha\gamma}, \quad \dot{y}_\gamma = -V_{\alpha\gamma}, \quad \dot{z}_\gamma = 0.$$

But
$$\dot{x}_\gamma^2 + \dot{y}_\gamma^2 + \dot{z}_\gamma^2 = V_{\gamma\beta}^2.$$

Hence
$$V_{\gamma\beta}^2 = V_{\alpha\gamma}^2 + V_{\alpha\beta}^2/\Omega_{\alpha\gamma}^2 \ldots \ldots \ldots (\text{iii}).$$

From (ii) and (iii) and (i) of $50 \cdot 3$

$$V_{\alpha\beta}^2/(1 - \Omega_{\alpha\beta}^{-2}) = V_{\alpha\gamma}^2/(1 - \Omega_{\alpha\gamma}^{-2}) \quad \ldots (\text{iv}).$$

$51 \cdot 3$ Evidently if δ be any other member of the collinear set of time-systems (α, β), then

$$V_{\alpha\delta}^2/(1 - \Omega_{\alpha\delta}^{-2}) = V_{\alpha\gamma}^2/(1 - \Omega_{\alpha\gamma}^{-2}) \ldots \ldots (\text{v}).$$

Hence if s be a collinear set of time-systems, and $\alpha, \beta, \delta, \epsilon$ be any four of its members,

$$V_{\alpha\beta}^2/(1 - \Omega_{\alpha\beta}^{-2}) = V_{\alpha\delta}^2/(1 - \Omega_{\alpha\delta}^{-2});$$

and hence, since $\Omega_{\alpha\delta} = \Omega_{\delta\alpha}$, we obtain

$$V_{\alpha\beta}^2/(1 - \Omega_{\alpha\beta}^{-2}) = V_{\delta\epsilon}^2/(1 - \Omega_{\delta\epsilon}^{-2}) = k_s \ldots (\text{vi}),$$

where k_s is a constant for the collinear set.

Furthermore, if γ be a time-system not belonging to s but related to α and s as explained in $51 \cdot 2$,

$$V_{\alpha\gamma}^2/(1 - \Omega_{\alpha\gamma}^{-2}) = k_s \quad \ldots \ldots (\text{vii}).$$

$51 \cdot 4$ Now let α, β, η be any three non-collinear time-systems, and construct a diagram to represent elements in the time-less space of α according to the familiar method of geometricians.

The points of the diagram symbolise α-points, and the straight lines of the diagram symbolise α-lines. Let O be any α-point and let OX be the direction in α-space of the velocity $V_{\alpha\beta}$. Then OX is the direction in α-space of the velocity (positive or negative) of any member of

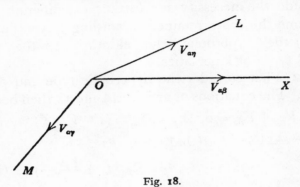

Fig. 18.

the collinear set (α, β). Let OL be the direction in α-space of the velocity $V_{\alpha\eta}$; by hypothesis OL is distinct from OX. Let OM be the α-line perpendicular to the α-plane LOX, and let γ be a time-system whose velocity in α, namely $V_{\alpha\gamma}$, is along OM. Let s denote the collinear set (α, β), s' the collinear set (α, γ), and s'' the collinear set (α, η). Hence from (vi) of 51·3

$$V_{\alpha\beta}^{2}/(1-\Omega_{\alpha\beta}^{-2})=k_{s}, \ V_{\alpha\gamma}^{2}/(1-\Omega_{\alpha\gamma}^{-2})=k_{s'}, \ V_{\alpha\eta}^{2}/(1-\Omega_{\alpha\eta}^{-2})=k_{s''}.$$

Hence from (vii) of 51·3

$$k_{s'}=k_{s}, \ k_{s'}=k_{s''}.$$

Thus
$$k_{s}=k_{s''} \ \dots\dots\dots\dots\dots(i).$$

Hence, since $V_{\alpha\beta}^{2}=V_{\beta\alpha}^{2}$ and $\Omega_{\alpha\beta}=\Omega_{\beta\alpha}$, it is easy to prove that k_{s} is the same for any pair of time-systems; in other words, that k_{s} is an absolute constant.

52. The Three Types of Kinematics. 52·1 There are thus three types of kinematics possible, according as

k_s is positive, negative, or infinite. The formally possible type where k_s is zero requires that either $\Omega_{\alpha\beta}$ or $V_{\alpha\beta}$ should be zero; by reference to (i) of 49·7 and to (i) of 51·1 this supposition is seen to lead to results in such obvious contradiction to experience as to preclude the necessity for further examination. Let us name the types retained (according to the familiar habit) the 'hyperbolic,' the 'elliptic' and the 'parabolic' types of kinematics.

52·2   First consider the hyperbolic type and put c^2 for k_s. The equations of articles 49 and 51 then become

$$V_{\alpha\beta} + V_{\beta\alpha} = 0, \quad \Omega_{\alpha\beta} = \Omega_{\beta\alpha} = (1 - V_{\alpha\beta}^2/c^2)^{-\frac{1}{2}} \ . \ . \text{(i)},$$

$$x_\beta = \Omega_{\alpha\beta}(x_\alpha - V_{\alpha\beta}t_\alpha), \quad y_\beta = y_\alpha, \quad z_\beta = z_\alpha,$$

$$t_\beta = \Omega_{\beta\alpha}\left(t_\alpha + \frac{V_{\beta\alpha}x_\alpha}{c^2}\right) \quad . \ . \text{(ii)},$$

$$\left. \begin{aligned}
\dot{x}_\beta &= (\dot{x}_\alpha - V_{\alpha\beta}) \Big/ \left(1 + \frac{V_{\beta\alpha}\dot{x}_\alpha}{c^2}\right) \\
\dot{y}_\beta &= \qquad \dot{y}_\alpha \Big/ \Omega_{\alpha\beta}\left(1 + \frac{V_{\beta\alpha}\dot{x}_\alpha}{c^2}\right) \\
\dot{z}_\beta &= \qquad \dot{z}_\alpha \Big/ \Omega_{\alpha\beta}\left(1 + \frac{V_{\beta\alpha}\dot{x}_\alpha}{c^2}\right)
\end{aligned} \right\} \quad \dots \text{(iii)}.$$

The equations of transformation, namely (ii), can be expressed symmetrically as between α and β by means of the scheme [where $i^2 = -1$]

	x_α	, y_α,	z_α,	ict_α
x_β	$\Omega_{\alpha\beta}$	, 0,	0,	$i\dfrac{V_{\alpha\beta}\Omega_{\alpha\beta}}{c}$
y_β	0	, 1,	0,	0
z_β	0	, 0,	1,	0
ict_β	$i\dfrac{V_{\beta\alpha}\Omega_{\beta\alpha}}{c}$,	0,	0,	$\Omega_{\beta\alpha}$

52·3 We notice that

$$\frac{\partial (x_\beta, y_\beta, z_\beta, t_\beta)}{\partial (x_a, y_a, z_a, t_a)} = 1 \dots\dots\dots(i).$$

The integral $\iiint dx_a\,dy_a\,dz_a\,dt_a$
taken throughout the four-dimensional region of the
set of event-particles which analyse [cf. 37·3] an event e
will be called the 'absolute extent' of e. It follows from
(i) that the absolute extent of an event is independent
of the time-system in which its measure is expressed.

Furthermore if K be any function of (x_a, y_a, z_a, t_a),
it can by (ii) of 52·2 be also expressed as a function of
$(x_\beta, y_\beta, z_\beta, t_\beta)$, and then by (i)

$$\iiint K\,dx_a\,dy_a\,dz_a\,dt_a = \iiint K\,dx_\beta\,dy_\beta\,dz_\beta\,dt_\beta\dots(ii),$$

or, in more familiar form,

$$\int dt_a . \iiint K\,dx_a\,dy_a\,dz_a = \int dt_\beta . \iiint K\,dx_\beta\,dy_\beta\,dz_\beta\dots(iii),$$

where the limits are taken to include some event.

We may expect important physical properties to be
expressible in terms of such integrals, in particular
where K is an invariant form for the equations of transformation of 52·2, and when the conditions, which the
quantity represented by the integral satisfies, are also
invariant in their expression in different time-systems.

The formulae of this subarticle hold of each type of
kinematics.

52·4 The hyperbolic type of kinematics has issued
in the formulae of the Larmor-Lorentz-Einstein theory
of electromagnetic relativity, namely, the theory by
which with a certain amount of interpretation the
electromagnetic equations are invariant for these
transformations.

The physical meaning of c is also well known;
namely, any velocity which in any time-system is of

magnitude c is of the same magnitude in every other time-system. No assumption of the existence of a velocity with this property or of the electromagnetic invariance has entered into the deduction of the kinematic equations of the hyperbolic type. A velocity greater than c cannot represent any time-system, and accordingly its physical significance must be entirely different from that of a velocity less than c.

52·5 It is easily proved from (ii) of 52·2 that

$$x_a^2 + y_a^2 + z_a^2 - c^2 t_a^2 = x_\beta^2 + y_\beta^2 + z_\beta^2 - c^2 t_\beta^2 \ldots \text{(i)}.$$

If the origin O and the event-particle P, i.e. (x_a, y_a, z_a, t_a), be co-momental and π be the time-system whose moment O_π contains P, then by (i)

$$x_a^2 + y_a^2 + z_a^2 - c^2 t_a^2 = x_\pi^2 + y_\pi^2 + z_\pi^2 \ldots \text{(ii)}.$$

If O and P be sequent and on a point-track, and π be the time-system whose point $OO_{\pi t}$ is occupied by P, then by (i)

$$c^2 t_a^2 - x_a^2 - y_a^2 - z_a^2 = c^2 t_\pi^2 \ldots \ldots \text{(iii)}.$$

Thus there are three ways in which the 'separation' between two event-particles (O and P) can be estimated; namely, (1) in any assumed time-system a the a-distance between the a-points occupied by the event-particles measures a-space separation: (2) the lapse of a-time between the a-moments occupied by the event-particles measures a-time separation: and (3) if the event-particles be co-momental, $\sqrt{(x_a^2 + y_a^2 + z_a^2 - c^2 t_a^2)}$ measures the 'proper' space separation and there is no 'proper' time separation; and if the particles be sequent,

$$\sqrt{\left\{ t_a^2 - \frac{x^2 + y^2 + z_a^2}{c^2} \right\}}$$

measures the 'proper' time separation and there is no 'proper' space separation.

In the framing of physical laws it is essential to consider what measure of separation is relevant. It is to be noted that there may be time-systems α (other than π) of special relevance to the phenomena in question. It is not at all obvious that invariance of form in respect to all time-systems is a requisite in the complete expression of such laws; namely, the demand for relativistic equations is only of limited applicability.

If O and P be on a null-track

$$x_a^2 + y_a^2 + z_a^2 - c^2 t_a^2 = 0 \dots \dots (\text{iv}).$$

Event-particles on the same null-track may be expected to have special physical relations to each other. Call such event-particles 'co-null.'

52·6 We may conceive a special time-system π associated (by some means) with each event-particle (x_a, y_a, z_a, t_a). Thus π is a function of these four co-ordinates of a particle; or in other words, $(\dot{x}_a, \dot{y}_a, \dot{z}_a)$ are functions of (x_a, y_a, z_a, t_a).

A correlation of time-systems to event-particles which is one-many, so that there is one and only one time-system corresponding to each event-particle, is called a 'complete kinematic correlation.' The portion of that correlation which only concerns event-particles at the time t_a is called a 'kinematic t_a-correlation.' Other portions can be selected by confining the event-particles to certain regions in the α-space.

If in a certain kinematic correlation the time-system π be correlated to (x_a, y_a, z_a, t_a), then π is called the time-system of (x_a, y_a, z_a, t_a) 'proper' to that correlation. The 'proper' time-system of an event-particle always refers to a certain kinematic correlation implicitly understood. Furthermore $(\dot{x}_a, \dot{y}_a, \dot{z}_a)$ is the velocity at (x_a, y_a, z_a) due to the implicitly understood kinematic correlation at the a-time t_a.

Then, π being the proper time-system at (x_a, y_a, z_a, t_a),

$$1 + \frac{V_{\beta a}\dot{x}_a}{c^2} = \Omega_{\beta\pi}/\Omega_{a\pi}\Omega_{a\beta} \dots\dots\dots \text{(i)}.$$

Then equations (iii) of 52·2 can be written

$$\Omega_{\beta\pi}\dot{x}_\beta = \Omega_{a\beta}\Omega_{a\pi}(\dot{x}_a - V_{a\beta}), \quad \Omega_{\beta\pi}\dot{y}_\beta = \Omega_{a\pi}\dot{y}_a,$$
$$\Omega_{\beta\pi}\dot{z}_\beta = \Omega_{a\pi}\dot{z}_a \dots \text{(ii)}.$$

The kinematic symmetry as between a and β is now apparent in the formulae. The first of equations (ii) can be replaced by

$$\frac{\dot{x}_a}{V_{a\beta}} + \frac{\dot{x}_\beta}{V_{\beta a}} + \frac{\dot{x}_a\dot{x}_\beta}{c^2} = 1 \dots\dots \text{(iii)}.$$

52·7 In considering the elliptic type of kinematics put $-h^2$ for k_s. The equations of article 51 are now embodied in the scheme

	x_a ,	y_a,	z_a,	$-ht_a$
x_β	$\Omega_{a\beta}$,	0,	0,	$\dfrac{V_{a\beta}\Omega_{a\beta}}{h}$
y_β	0 ,	1,	0,	0
z_β	0 ,	0,	1,	0
$-ht_\beta$	$-\dfrac{V_{\beta a}\Omega_{\beta a}}{h}$,	0,	0,	$\Omega_{\beta a}$

Also $V_{a\beta} + V_{\beta a} = 0$, $\Omega_{a\beta} = \Omega_{\beta a} = (1 + V_{a\beta}^2/h^2)^{-\frac{1}{2}}$..(i),

and $\dot{x}_\beta = \dfrac{\dot{x}_a - V_{a\beta}}{1 - \dfrac{V_{\beta a}\dot{x}_a}{h^2}}, \quad \dot{y}_\beta = \dfrac{\dot{y}_a}{\Omega_{a\beta}\left(1 - \dfrac{V_{\beta a}\dot{x}_a}{h^2}\right)},$

$$\dot{z}_\beta = \frac{\dot{z}_a}{\Omega_{a\beta}\left(1 - \dfrac{V_{\beta a}\dot{x}_a}{h^2}\right)} \dots\dots \text{(ii)}.$$

Also $x_a^2 + y_a^2 + z_a^2 + h^2 t_a^2 = x_\beta^2 + y_\beta^2 + z_\beta^2 + h^2 t_\beta^2$ ..(iii).

The fundamental distinction between space and time, i.e. between rects and point-tracks, has failed to find any expression in the formulae for measurement relations. Accordingly with this type of kinematics, it would be natural to suppose that the distinction does not exist and that every rect was a point-track and every point-track a rect. This conception is logically possible but does not appear to correspond to the properties of the external world of events as we know it. Furthermore the electromagnetic equations lose their invariant property.

Altogether there appear to be good reasons for putting aside the elliptic type of kinematics as inapplicable to nature.

52·8 In the parabolic type of kinematics we put $k_s = \infty$. Hence

$$\Omega_{\alpha\beta} = 1 \dots\dots\dots\dots\dots\text{(i)}.$$

Then from (ii) of 51·1 and (ii) of 50·3 and (iii) of 50·1

$$\Omega_{\alpha\beta}''' = 0, \quad \Omega_{\alpha\beta}'' = 1, \quad \Omega_{\alpha\beta}' = -V_{\alpha\beta} \ \dots\text{(ii)}.$$

Thus equations (i) of 49·7 give

$$x_\beta = x_a - V_{\alpha\beta} t_a, \ y_\beta = y_a, \ z_\beta = z_a, \ t_\beta = t_a ..\text{(iii)}.$$

These are the formulae for the ordinary Newtonian relativity.

These formulae are well in accordance with common sense and are in fact the formulae naturally suggested by ordinary experience. To some extent the hyperbolic formulae lead to unexpected results, though, if c be a velocity not less than that of light, the divergences from the deliverances of common sense take place in respect to phenomena which are not manifest in ordinary experience. But when by refined methods of observation the divergences between the two types of kinematics should be apparent to the senses, experiment has, so far, pronounced in favour of the hyperbolic type. Accordingly it is this type which we consider in the sequel.

52·9 There is however one objection to the hyperbolic type, as compared to the parabolic type, which is worth considering. In the hyperbolic kinematics there is an absolute velocity c with special properties in nature. The difficulty which is thus occasioned is rather an offence to philosophic instincts than a logical puzzle. But certainly our familiar experience, in some way which it is difficult to formulate in words, leads us to shun the introduction of such absolute physical quantities. This particular difficulty is largely diminished by noting that the existence of c with its peculiar properties really means that the space-units and time-units are comparable; namely, there is a natural relation between them to be expressed by taking c to be unity. Either the time-unit would then be inconveniently small or the space-unit inconveniently large; but this inconvenience does not alter the fact that congruence between time and space is definable. Always when a possible definition of congruence is omitted, such absolute physical quantities occur. The fact that, so far as time and space are concerned, the existence of a congruence theory seems paradoxical is due to absence of any phenomena depending on that theory except in very exceptional circumstances produced by refined observations.

PART IV

THE THEORY OF OBJECTS

CHAPTER XIV

THE LOCATION OF OBJECTS

53. Location. 53·1 We conceive objects as located in space. This conception of location in space is distinct from that of being situated in an event, though the two concepts are closely allied by a determinate connection. The notion of the situation of an object is logically indefinable being one of the ultimate data of science; the notion of the location of an object is definable in terms of the notion of its situation.

An object is said to be 'located' in an abstractive element if there is a simple abstractive class 'converging' to the element and such that each of its members is a situation of the object.

In general when an object is located in an abstractive element there will be many simple abstractive classes converging to the element and such that each of their members is a situation of the object. In any specific case of location usually all abstractive classes of a certain type will possess the required property.

It follows from this definition that, in the primary signification of location, an object is located in an element of instantaneous space. The notion of location in an element of time-less space follows derivatively by correlating the elements of instantaneous space to the elements of time-less space in the way already described. In our immediate thoughts which follow perception we make a jump from the situation of an object within the short specious present to its location in instantaneous space, and thence by further reflexion to its location in time-less space. Thus location in space is always an ideal of thought and never a fact of perception. An object may be located in a volume, an area, a route, or an event-particle of instantaneous space, and thence derivatively it will be located in a volume, or an area, or a segment, or a point of time-less space.

53·2 In considering the scientific object it is the occupied event which corresponds to the situation of the physical object. The occupied event is the situation of the charge, in so far as the single scientific object is conceived as an (ideal) physical object.

53·3 There are evidently many different kinds of location which satisfy the general definition of location in an abstractive element, even when the kind of abstractive element is assigned. These differences mainly arise from differences in the relations of objects to parts of their situations. An object is an atomic entity and as such is related to its situations. But a situation is an event with parts of various kinds, and we have to consider the various kinds of relationships which objects may have to various kinds of parts of their situations.

For example, if the sense-object 'redness, of a definite shade' be located in an area, it will be located in any portion of that area; and this arises from the fact that if it be situated in an event, it is also situated in any portion of that event. But it is not true that if a chair be situated in an event, that the chair—as one atomic object— is situated in any part of the event though it is so situated in some parts. Again a tune cannot be situated in any event comprised in a duration too short for the successive notes to be sounded. Thus for a tune a minimum quantum of time is necessary.

54. Uniform Objects. 54·1 It will be convenient to classify objects according as they do or do not satisfy certain important conditions respecting their relations to their situations.

'Uniform' objects are objects with a certain smoothness in their temporal relations, so that they require no minimum quantum of time-lapse in the events which are their situations. These are objects which can be said to exist 'at a given moment.' For example, a tune is not an uniform object; but a chair, as ordinarily recognised, is such an object. The example of the chair, and the dissolution of its continuous materials with specific physical constants into assemblages of electrons, warn us that a problem remains over for discussion after we shall have defined the meaning to be assigned to 'uniformity.'

54·2 In order to explain more precisely the theory of uniform objects, it is convenient to make a few definitions:

A 'slice' of an event e in a time-system a is that part of e lying between two moments of a, where both moments intersect e. The two moments are called the terminal moments of the slice, and the volumes in which the terminal moments intersect e are called the terminal volumes. For brevity a slice of e in the time-system a is called an 'a-slice of e.'

It follows from the continuity of events that any a-moment lying between the terminal moments of an a-slice of e intersects e in a volume. Such a volume is called an α-section of the slice. A slice is itself an event which stretches throughout the duration bounded by its terminal moments. Thus if the duration be the specious present for some percipient, the slice of e is the part of the event e which falls within that specious present.

54·3 The properties of uniform objects will be enunciated as a set of laws regulating their character.

Law I. If α be any time-system and e be a situation of an uniform object O, then an α-slice of e exists which is a situation of O.

Law II. If α be any time-system and e be a situation of an uniform object O and e' be an α-slice of e which is a situation of O, then every α-slice of e' is a situation of O.

Law I can roughly be construed as meaning that if an uniform object O has been situated in any event, then there is some period of time (in any time-system) during which it has existed; and in the same way Law II means that if an uniform object has existed during any period of time, then it has existed during any shorter period within that period. These laws are obvious as applied to uniform objects, but not so obvious for objects in general, as 'object' is here defined. For example a musical note cannot exist in a period of time shorter than its period of vibration, and a percipient whose specious present was too short could not hear it. It follows from law II that if an uniform object O is situated in an event e and e' be an α-slice of e which is a situation of O, then an abstractive class of α-slices converging to any α-section of e' can be found such that O is situated in each member of the class. Hence evidently O is located in every α-section of e'. This is the conception of an uniform object being located in a spatial volume at a durationless moment of time.

With certain explanations and limitations laws I and II apply to many types of objects. In fact it requires an effort to realise that there are cases to which they do not apply. They have been stated above in the most formal manner to exhibit the fact that, when they do apply, they are empirical laws of nature and not *à priori* logical truths.

55. Components of Objects. 55·1 The concept of a 'component' of a main object is difficult to make precise. A component of an object O is another distinct object O' such that (i) whenever O is situated in an event e, there is an event e', which is either e itself or a part of e, in which O' is situated, and (ii) O' may also be situated in an event e'' which is not a situation of O or any part of a situation of O.

Thus a component is necessary to the main object, but the main object is not necessary to the component. For example, a certain note may be necessary for a certain tune, but the note can be sounded without the tune. The main object requires its component, but the component does not require the main object.

But this general idea of a component is not of great importance apart from further specialisation. There are many such specialisations; but in science there are three which are of peculiar importance, namely, 'concurrent components,' 'extensive components' and 'causal components.'

55·2 An object O' is a 'concurrent' component of an object O when it is a component of O, and if e be any situation of O, there is an event e' which is part of e and is such that (i) it is a situation of O' and (ii) it is cut in a slice which is a situation of O' by any duration which cuts e in a slice which is a situation of O.

Thus a concurrent component lasts concurrently with the main object in any time-system.

CHAPTER XV

MATERIAL OBJECTS

56. Material Objects. 56·1 A material object is essentially a material object of a certain definite sort; namely, we define sorts of material objects, which are sets of objects with certain definite peculiarities, and a material object is such because it is a member of one of these sorts. For example a piece of wood is a material object because it belongs to the class of wooden objects and because this class possesses the requisite peculiarities. Similarly a charge of electricity is a material object for an analogous reason.

The objects which compose a set (μ) form a sort of 'material' objects when (i) the objects of the set μ are all uniform, (ii) not more than one member of μ can be located in any volume, (iii) no member of μ can be located in two volumes of the same moment, (iv) if O_1 and O_2 be two members of μ respectively located in non-overlapping volumes in the same moment, then any pair of situations of O_1 and O_2 respectively are separated events, (v) if O be a member of μ situated in an event e, and located in the volume V which is a section of e, and V_1 be any volume which is a portion of V, then there is a member of μ which is located in V_1 and is a concurrent component of O.

56·2 If O be a material object of a certain sort and V be a volume in which O is located and V_1 be a portion of V, then the material object of the same sort as O which is located in V_1 is called an 'extensive component' of O.

56·3 It is by means of the properties of material
objects that the atomic properties of objects are combined in mathematical calculations with the extensive
continuity of events. Apart from material objects
mathematical physics as at present developed would be
impossible. For example where the physicist sees the
electron as an atomic whole, the mathematician sees
a distribution of electricity continuous in time and in
space and capable of division into component objects
which are also analogous distributions.

57. Stationary Events. 57·1 In order to understand
the theory of the motion of material objects, it is first
necessary to define the concept of a 'stationary' event.
Consider some given time-system π, and let V denote
a volume lying in a certain moment M of this time-
system. Let d be a duration of π bounded by moments
M_1 and M_2, and inhered in by M; so that M_1, M_2, M
are three parallel moments of the time-system π, and
M lies between M_1 and M_2. The volume V is the locus
of a set of event-particles and each of these event-
particles lies in one and only one station of the duration
d. Also each station of d either does not intersect V or
intersects it in one event-particle only. The assemblage
of event-particles lying on stations of d which intersect
V [namely, each event-particle lying on one of these
stations] is the complete set of event-particles analysing*
an event. Such an event is called stationary in the time-
system π and stretches throughout the duration d. It
can also be called 'stationary in d,' since d defines the
time-system π. Every event-particle within the event
lies on a station of d; and a station of d either has all
its event-particles lying within the event or none of

* Cf. subarticle 37·3, Chapter X, Part III.

them. The volume V is the section of the event by the moment M. Furthermore if M' be any other moment of the time-system π lying between M_1 and M_2, it intersects the event in a volume V' which is a geometrical replica of the volume V. The moments M_1 and M_2 which bound the duration d are the terminal moments of any event which is stationary in d. The stations of d lying in the event intersect M_1 and M_2 in terminal volumes V_1 and V_2 which are geometrical replicas of V and V'. A volume, such as V', in which a moment of π intersects an event stationary in π is called a 'normal cross-section' of the event. A moment of another time-system a which intersects the stationary event in a volume U, but does not intersect either of the terminal volumes, is said to intersect it in an 'oblique cross-section.' All the oblique cross-sections of a stationary event which are made by moments of the same time-system are geometrical replicas of each other.

57·2 Consider an event e stationary in the time-system π, and let a be another time-system. Let v_π be the measure of the normal cross-sections of e and let v_a be the measure of the oblique cross-sections made by moments of a. We require the ratio of v_a to v_π. Take (as usual) mutual axes for π and a, and let the event-particle which is the origin lie in M_1 which is the antecedent terminal moment of e. Then M_1 is at the π-time zero, and let M_2 (the subsequent terminal moment) be at the π-time t_π. Then if $(x_\pi, y_\pi, z_\pi, 0)$ be the π-coordinates of the event-particle in which a station s (of the set composing the event e) intersects M_1, the π-coordinates of the other end (the subsequent end) of s in M_2 are $(x_\pi, y_\pi, z_\pi, t_\pi)$.

Furthermore let (x_a, y_a, z_a, t_a) be the a-coordinates of the antecedent end of s, and let (x'_a, y'_a, z'_a, t'_a) be the a-coordinates of the subsequent end of s. Then by the usual formulae [cf. subarticle 52·2]

$$x_a = \Omega_{a\pi} x_\pi, \ y_a = y_\pi, \ z_a = z_\pi, \ t_a = \Omega_{a\pi} V_{a\pi} x_\pi / c^2,$$

and $x'_a = \Omega_{a\pi}(x_\pi - V_{\pi a} t_\pi), \ y'_a = y_\pi, \ z'_a = z_\pi,$

$$t'_a = \Omega_{a\pi}\left(t_\pi + \frac{V_{a\pi} x_\pi}{c^2}\right).$$

Hence $t'_a - t_a = \Omega_{a\pi} t_\pi.$

But, by analogous reasoning to that for the elementary case of geometrical parallelograms, the absolute extent of the event e can be expressed as $v_\pi t_\pi$ and as $v_a (t'_a - t_a)$. Hence $v_a (t'_a - t_a) = v_\pi t_\pi.$

Thus $v_a = \Omega_{a\pi}^{-1} v_\pi \ldots \ldots \ldots \ldots (1).$

57·3 The stations of a duration d of a time-system π are portions of points of the time-less space of π [the π-space].

Thus by prolonging the stations which constitute the stationary event e we obtain the assemblage of π-points which is the complete assemblage of π-points intersecting the cross-sections of e, each event-particle in each cross-section lying on one and only one such π-point and each of these π-points intersecting each cross-section in one event-particle. The assemblage of these π-points is a volume of the π-space, and the successive instantaneous volumes which are the normal cross-sections of e [stationary in π] each occupy this same volume in the π-space. Thus the stationary event e during the lapse of π-time throughout which it endures is happening at the same place in the π-space.

But the successive oblique cross-sections of e formed by moments of another time-system a are instantaneous volumes which successively occupy different volumes in the α-space. These instantaneous volumes travel in the α-space, sweeping over it with the uniform velocity $V_{a\pi}$, namely the velocity due to the time-system π in the α-space.

57·4 A 'normal slice' of a stationary event is the slice of it cut off between any two normal cross-sections. An 'oblique slice' of a stationary event is the slice of it cut off between any two parallel oblique cross-sections. A normal slice of a stationary event is itself a stationary event in the same time-system.

58. *Motion of Objects.* 58·1 A material object is 'motionless' within a duration when throughout that duration the material object and its extensive components are all situated in stationary events.

In the case of a motionless material object, Law I for uniform objects can be made more precise, as follows:

If O be a material object motionless in the duration d and e be the stationary event extending throughout d in which it is situated, then O is situated in any oblique slice of e.

The accompanying figures illustrate (i) the kind of slice which is included in this law and (ii) the kind of slice which is excluded.

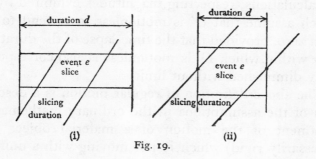

Fig. 19.

It immediately follows that—with the nomenclature of the enunciation of the law—O is located in every oblique cross-section of e.

If π be the time-system of the duration in which O is motionless and, in some other time-system a, d' be the duration of maximum extent which intersects e in an oblique slice, then throughout d' in the time-less space of a the material object O has a uniform motion of translation with the velocity of π in a.

58·2 This property, possessed by a material object which is motionless in a time-system π, of being situated in every oblique slice of its stationary situation is a fundamental physical law of nature. Namely, percipients cogredient with different time-systems can 'recognise' the same material objects. In other words, the character of a material object is not altered by its motion.

58·3 The motion of a material object O is 'regular' when if V be any volume in which it is located and P be any event-particle in V, and V' be any variable volume which contains P and is a portion of V, and O' be the extensive component of O which is located in V', then, as V' is progressively diminished without limit, a time-system π can be found such that the errors of calculations, respecting magnitudes exhibited by O' which assume that O' is motionless in π, tend to the limit zero, provided that the time-lapse of the durations in π within which O' is motionless is also correspondingly diminished without limit.

The above definition of regular motion is a description of the assumptions in the ordinary mathematical treatment of the motion of a material object (not necessarily rigid) which is not moving with a uniform motion of translation. If α be the standard time-system to which motions are referred, then the velocity of π in α is the velocity at the event-particle P [i.e. at the α-space point p_α at the α-time t_α] of the material object.

59. Extensive Magnitude. 59·1 A theory of extensive magnitude is required to complete the theory of material objects.

Let O and O' be two objects (material or otherwise), then the statement that O and O' possess quantities of a certain kind and that the ratio of the quantity O to the quantity O' has a certain definite numerical value is a reference to some determinate method of comparison of O to O' which is the defining characteristic of that kind of quantity*.

The quantity of a certain kind possessed by a material object O is called 'extensive' when it is a determinate function of the quantities of the same kind possessed by any two of its extensive components which (i) are exhaustive of O and (ii) are non-overlapping [i.e. have no extensive component in common].

If the determinate function be that of simple addition [so that, q, q_1, q_2 being the quantities possessed respectively by O and its two extensive components,
$$q = q_1 + q_2],$$
then the kind of quantity will be called 'absolutely' extensive. When an extensive quantity is not absolutely extensive, it will be called 'semi-extensive.'

59·2 It is usual in philosophical discussions to confine the term 'extensive quantity' to what is here defined as 'absolutely extensive quantity,' and to ignore entirely the occurrence of semi-extensive quantities. But in physical science semi-extensive quantities are well

* Cf. *Principia Mathematica.*

known. For example, consider a sphere of radius a uniformly charged with electricity throughout its volume. Divide the sphere into two parts, namely a concentric nucleus of radius c and a shell of thickness $a - c$. Then the electromagnetic mass of the whole sphere is not the sum of the electromagnetic masses of these two parts, but is to be calculated by a quadratic law from the charges.

A material object expresses the spatial distribution of a quantity of 'material,' when the quantity is absolutely extensive.

59·3 The volume-density, at a time t_a in the a-space of a time-system a, of the distribution of any absolutely extensive quantity possessed by a material object O is calculated by the ordinary mathematical formula. Consider any event-particle P occupying the a-point p_a at the a-time t_a. Let dv_a be the measure of a volume in the a-space which contains p_a; and let O' be the extensive component of O located in dv_a, if there be such an extensive component. Let q be the measure of the quantity possessed by O'. Then the limit of the ratio of q to dv_a, as dv_a is indefinitely diminished, is the density at p_a at the time t_a of the material [i.e. of the absolutely extensive quantity].

59·4 The above definitions contemplate quantities immediately possessed by the extensive objects as such, for example, charges of electricity and intensities of sense-objects. But there are also quantities which are only mediately possessed by the objects, but are immediately possessed by the events which are their situations. Such quantities may vary with the variation in the situation of the object mediately possessing them.

A mediately possessed quantity may for a certain

type of material objects satisfy the characteristic condition for an extensive quantity. In that case it is an extensive quantity mediately possessed by that type of material objects. All variable extensive quantities are of this mediate character. A quantity mediately possessed by a material object O at a moment M_a [i.e. at a time t_a] of a time-system a is the limit of the quantity possessed by the successive converging situations of O in the successive durations of an abstractive class (of durations in the time-system a) which converges to M_a.

The volume-density, at a time t_a in the a-space of a time-system a, of the distribution of any absolutely extensive quantity mediately possessed by a material object O is calculated according to the preceding definition for the case of immediately possessed quantities, except that the 'quantity mediately possessed by O (or by an extensive component of O) *at the time t_a*' must be substituted everywhere for the 'quantity possessed by O (or by an extensive component of O).'

59·5 We can compare the volume-densities ρ_a and ρ_β of an absolutely extensive quantity for two time-systems a and β respectively at a given event-particle P, assuming, as we may assume, that the motion of the material object possessing (mediately or immediately) the quantity is regular.

Let π be the time-system in which the object O is stationary at P, and let ρ_π be the volume-density at P for the time-system π. Let M_a, M_β, M_π be the moments in a, β, and π respectively which contain P. Let dv_π be the measure of a small volume in M_π which contains P [and therefore the measure of the volume in the timeless π-space which this instantaneous volume occupies]. Consider the event (e_π) stationary in π of which this small volume dv_π is a normal cross-section, and bounded by terminal moments M'_π and M''_π on either side of M_π and both near M_π. Then, by the theory of regular motion, we can take this stationary event (e_π) as the situation of an extensive component of O, when dv_π is small enough and the duration bounded by M'_π and M''_π is short enough. Let dv_a and dv_β be the measures of the volumes which are the oblique cross-sections of e_π made by M_a and M_β. Then ultimately $\rho_\pi dv_\pi$, $\rho_a dv_a$, and $\rho_\beta dv_\beta$ are expressions for the measure of the quantity possessed by O.

Hence $\rho_a dv_a = \rho_\beta dv_\beta = \rho_\pi dv_\pi.$

But by equation (1) of 57·2 of this chapter,
$$dv_a = \Omega_{a\pi}^{-1} dv_\pi, \quad dv_\beta = \Omega_{\beta\pi}^{-1} dv_\pi.$$

Thus $\rho_a \Omega_{a\pi}^{-1} = \rho_\beta \Omega_{\beta\pi}^{-1} = \rho_\pi \ldots\ldots\ldots(2).$

Now take the mutual axes for a and β, and let (x_a, y_a, z_a, t_a) and $(x_\beta, y_\beta, z_\beta, t_\beta)$ be the coordinates of P in a and β respectively, and let $(\dot{x}_a, \dot{y}_a, \dot{z}_a)$ and $(\dot{x}_\beta, \dot{y}_\beta, \dot{z}_\beta)$ be the velocities due to π in a and β respectively. Then by equation (i) of 52·6,

$$\rho_\beta = \rho_a \Omega_{a\beta}\left(1 - \frac{V_{a\beta}\dot{x}_a}{c^2}\right)\ldots\ldots\ldots(3).$$

59·6 Now let $\dfrac{d}{dt_a}$ denote differentiation following the motion $(\dot{x}_a, \dot{y}_a, \dot{z}_a)$ at (x_a, y_a, z_a, t_a), and let $\dfrac{\partial}{\partial t_a}$ denote differentiation at the point (x_a, y_a, z_a).

Then it is easily proved that
$$\Omega_{a\pi}\left\{\frac{1}{\rho_a}\frac{d\rho_a}{dt_a} + \operatorname{div}_a(\dot{x}_a, \dot{y}_a, \dot{z}_a)\right\}$$
$$= \frac{\Omega_{a\pi}^2}{\rho_a}\frac{d(\rho_a\Omega_{a\pi}^{-1})}{dt_a} + \left\{\frac{\partial\Omega_{a\pi}}{\partial t_a} + \operatorname{div}_a(\Omega_{a\pi}\dot{x}_a, \Omega_{a\pi}\dot{y}_a, \Omega_{a\pi}\dot{z}_a)\right\}.$$

Also
$$\Omega_{\alpha\pi} \frac{d}{dt_\alpha} \doteq \Omega_{\beta\pi} \frac{d}{dt_\beta}.$$

Hence from equation (2) of 59·5 above
$$\frac{\Omega_{\alpha\pi}^2}{\rho_\alpha} \frac{d\,(\rho_\alpha \Omega_{\alpha\pi}^{-1})}{dt_\alpha} = \frac{\Omega_{\beta\pi}^2}{\rho_\beta} \frac{d\,(\rho_\beta \Omega_{\beta\pi}^{-1})}{dt_\beta}.$$

Again by using the formulae of article 52, we can prove that
$$\frac{\partial \Omega_{\alpha\pi}}{\partial t_\alpha} + \mathrm{div}_\alpha\,(\Omega_{\alpha\pi}\dot{x}_\alpha,\; \Omega_{\alpha\pi}\dot{y}_\alpha,\; \Omega_{\alpha\pi}\dot{z}_\alpha)$$
$$= \frac{\partial \Omega_{\beta\pi}}{\partial t_\beta} + \mathrm{div}_\beta\,(\Omega_{\beta\pi}\dot{x}_\beta,\; \Omega_{\beta\pi}\dot{y}_\beta,\; \Omega_{\beta\pi}\dot{z}_\beta).$$

From these results we immediately deduce
$$\Omega_{\alpha\pi} \left\{ \frac{1}{\rho_\alpha} \frac{d\rho_\alpha}{dt_\alpha} + \mathrm{div}_\alpha\,(\dot{x}_\alpha,\, \dot{y}_\alpha,\, \dot{z}_\alpha) \right\}$$
$$= \Omega_{\beta\pi} \left\{ \frac{1}{\rho_\beta} \frac{d\rho_\beta}{dt_\beta} + \mathrm{div}_\beta\,(\dot{x}_\beta,\, \dot{y}_\beta,\, \dot{z}_\beta) \right\}. \quad ..(4).$$

Now the condition that the total extensive quantity which is the 'charge' of any extensive component never varies when conceived as distributed through the α-space is
$$\frac{1}{\rho_\alpha} \frac{d\rho_\alpha}{dt_\alpha} + \mathrm{div}_\alpha\,(\dot{x}_\alpha,\, \dot{y}_\alpha,\, \dot{z}_\alpha) = 0.$$

This is the well-known equation of continuity. Now equation (4) shows that if this equation holds for the space of any time-system, it holds for the spaces of all time-systems.

When the equation of continuity holds, the 'charge' of any extensive component of the material object under consideration never varies. Hence it is a mere matter of words and definition whether the charge is said to be mediately possessed by the object or immediately possessed.

CHAPTER XVI

CAUSAL COMPONENTS

60. Apparent and Causal Characters. 60·1 Are there any material objects in nature? That there are such bodies is certainly an assumption habitually made in the applications of mathematics. But the assumption does not supersede the necessity for enquiry.

We may roughly summarise the properties of material objects, as here defined, by saying that they should be continuous both in time and in space. But this is just what ordinary perceptual objects appear to be. Now perceptual objects are what they appear to be; for a perceptual object is nothing else than the permanent property of its situations, that they all shall exhibit those appearances. Accordingly if a perceptual object appears to be a material object, it is a material object.

Now here a difficulty arises; for we all know that, according to Dalton's atomic theory of chemistry, any apparently continuous substance is a discrete collection of molecules, and that furthermore, according to the more recent theories, a molecule is a discrete collection of electric charges. Accordingly, as we are told, if we could take the minutest drop of water and magnify it, the phenomena would be analogous to those of a swarm of flies in a room.

It would appear therefore that we are mistaken in classifying a drop of water as being a material object.

60·2 The position that we seem to have arrived at is that on the one hand a drop of water is a material object, because it appears to be one and it is whatever it appears to be, and that on the other hand it is really something quite different.

Such paradoxes mean that vital distinctions have been overlooked. We must distinguish between the drop of water as it appears, the event which is its situation, and the character of the event which causes the event to present that appearance. Namely, there is the appearance of the drop of water. This is character No. 1 of the event, the apparent character, and is a material object. Again there is the character of the event which is the cause of character No. 1. This is character No. 2 of the event and is its causal character. According to the doctrine of science, character No. 2 is not a material object.

60·3 But why trouble about causal characters? What has pushed science into their consideration? The impelling reason is the complex bewildering relationships of the apparent characters. Apparent characters essentially involve reference to percipient events, and may be very trivial qualities of the events which they characterise. For example, all delusive perceptual objects are apparent characters of events.

In the case of a delusive perceptual object character No. 2 of its situation has no existence, except so far as the event is necessarily still a 'passive condition' according to the nomenclature of Chapter VII of Part II. The active conditioning events for a delusive perceptual object must be sought elsewhere than in its situation. Let us confine ourselves to the consideration of non-delusive perceptual objects, that is, to physical objects.

60·4 But the line of separation between delusive and non-delusive perceptual objects is not quite so clear as we might wish. The definition of delusiveness and non-delusiveness is sufficiently obvious, namely, a perceptual object is non-delusive when it is the apparent character of an event which is itself an active condition for the appearance of that character as perceptible from all percipient events. In the nomenclature of Chapter VII of Part II the situation of a physical object is its 'generating event.'

Now if this definition is to be taken to the foot of the letter, all perceptual objects are delusive; for all perception is belated. The sun which we see is the apparent character of an event simultaneous with our percipient event, and this event is about eight minutes subsequent to the generating event corresponding to that appearance of the sun. In the case of other astronomical phenomena the discrepancy is more glaring. In the case of terrestrial perceptual objects the discrepancy is less glaring in many cases, though for sounds it is very insistent and is the reason of their very indeterminate situations. But, speaking generally and admitting exceptions, for the greater part of ordinary domestic perceptions the belatedness of the apparently characterised event behind the causally characterised event is a small fraction of the percipient's specious present.

Accordingly our knowledge of causal characters is a theory built up by ignoring this element of delusiveness in all perceptual objects, and then by introducing it as an additional correction in the exceptional cases.

61. Transition from Appearance to Cause. 61·1 But how can we pass from appearance to cause, seeing that our knowledge of nature is confined to awareness of appearance? For example, all measurement is a measurement of appearance.

Evidently therefore causal characters can only be directly known to us as functions of apparent characters. They are characters of characters. For example, a quantity which we assign to a physical object as the result of some measurement is a character of its apparent character.

61·2 It is necessary however to avoid a misunderstanding; the causal character of an event is not merely a function of the apparent character of that event. It is in truth a function of the apparent characters of all events, though in general the apparent character of that event—or of an associated event of somewhat later date—is the dominant element in the formation of the function. For example, a quantity determined by measurement is a relation of the apparent character of the event to the apparent characters of other events. But it is the dominance of the apparent character which in practice makes the discovery of the causal character generally possible; for it assigns the situation of the causal character. This dominance is merely a practical aid to the discovery of causal characters and has in it no element of necessity. Indeed as causal characters are progressively discovered, scientific theory assigns causal characters to events which are destitute of apparent character—namely the events forming the ether in empty space and time.

61·3 So far the explanation of causal characters has exhibited them as the outcome and issue from apparent characters, whereas the causal idea, which is that of science, requires the causal characters should be the origin of the apparent characters. We have to seek the reason for this inversion of ideas.

Causal characters are much simpler than apparent

characters; are more permanent than apparent characters; and depend almost entirely on the event itself,
involving other events only (in general) as passive conditions providing the necessary background of a whole
continuum of nature. The climb from the sense-object
to the perceptual object, and from the perceptual object
to the scientific object, and from the complex scientific
object (such as the molecule) to the (temporally, in a
stage of science) ultimate scientific object (such as the
electron) is a steady pursuit of simplicity, permanence,
and self-sufficiency, combined with the essential attri-
bute of adequacy for the purpose of defining the
apparent characters.

61·4 The relations of sense-objects to their situations
are complex in the extreme, requiring reference to
percipient events and transmitting events. Apart from
some discovery of laws of nature regulating the associations of sense-objects, it is impossible by unintelligent
unsorted perception to form any concept of the character of an event from the sense-objects which might be
situated there for percipients suffering from any normal
or abnormal perceptions.

The first stage is the discovery of perceptual objects.
These objects are first known by the instinctive 'conveyance' of abnormal perceptions of sense-objects associated with normally perceived sense-objects. The test
of alternative possibilities of normal perception and the
discovery of a permanent character in the association
which can be expressed independently of any particular
percipient event decides between delusive perceptual
objects and physical objects.

61·5 The introduction of physical objects enables
us in considering the characters of events to sweep

aside the boundless eccentricities of abnormal perceptions. We are still at the stage of apparent characters,
but rules have been attained, either by instinctive
practice or by the exercise of intelligence or by the
interplay between the two, by which we know what to
attend to and what to discard in judging the character
of an event from the situations of sense-objects. A
physical object is the apparent character of its situation.
Physical objects are found to be 'material' objects.

61·6 Science now intervenes with the express purpose
of exhibiting our perceptions as our awareness of the
characters of events and of relations between characters
of events. All perceptions are included in the scope
of this aim of science, namely, including abnormally
perceived sense-objects and delusive perceptual objects.

61·7 The origin of the concept of causation (in this
application of the term) is now manifest. It is that
of the part explaining the whole—or, avoiding this un-
technical use of 'part' and 'whole,' it is that of some
explaining all. For the physical objects were obtained by
discarding abnormalities, and physical objects express
the characters of events, and all our perceptions (including abnormalities) arise from awareness of these
characters.

61·8 But physical objects fail to satisfy the requirements of science. They lack definiteness and permanence, and are not adequate for the purposes of
explanation. Now the characters of their mutual relations disclose further permanences recognisable in
events and among these are the scientific objects. The
gradual recognition of these permanences was at first
the slow product of civilised thought without conscious
direction. As regards their conscious discovery various

stages may be discerned in scientific history, which sum up the previous growth of ideas and initiate new epochs. One stage is marked by Archimedes' discovery of specific gravity, and another by Newton's discovery of mass. The simplicity of what, in its relation to appearance, is so abstract was then beginning to be discovered, and also its permanence and self-sufficiency as a quality of events. A third stage is the introduction of the concept of molecules and atoms by Dalton's atomic theory. Finally there arose the concepts respecting the ether, which we here construe as meaning the concept of events in space empty of appearances.

61·9 These causal characters, which are the characters of apparent characters, are found to be expressible as certain scientific objects, molecules and electrons, and as certain characters of events which do not necessarily themselves exhibit any apparent characters. If we follow the route of the derivation of knowledge from the intellectual analysis of sensible experience, molecules and electrons are the last stage in a series of abstractions. But a fact in nature has nothing to do with the logical derivation of concepts. The concepts represent our abstract intellectual apprehension of certain permanent characters of events, just as our perception of sense-objects is our awareness of qualities of nature resulting from the shifting relations of these characters. Thus scientific objects are the concrete causal characters, though we arrive at them by a route of apprehension which is a process of abstraction. In the same way, what, in the form of a sense-object, is concrete for our awareness, is abstract in its character of a complex of relations between scientific objects. Thus what is concrete as causal is abstract in its derivation from the apparent, and what is concrete as apparent is abstract in its derivation from the causal.

The ultimate scientific objects (at present, electrons and positive electric charges) are 'uniform' objects; and, in the limited sense of charges in the 'occupied' events, they are also 'material' objects. There does not appear to be any reason, other than the very natural desire for simplicity, for the assumption that ultimate scientific objects are uniform. Some of the atomic and 'quantum' properties of nature may find their explanation in the assumption of non-uniform ultimate scientific objects which would introduce the necessary discontinuities.

61·91 The causal character of the situation of a physical object is the fact that this situation contains a certain assemblage of ultimate scientific objects; namely, the fact that among the parts of this situation are various parts which are the occupied events of these scientific objects. The 'causal components' of a physical object are the scientific objects which occupy parts of the situation of the physical object, and whose total assemblage is what constitutes the qualities which are the apparent character which is the physical object apparent in the situation.

61·92 An adjustment, ordinarily negligible but often important, has to be made to allow for the belatedness of perception. Two situations are thus involved (even although in ordinary cases they are practically identical), namely the situation of the physical object from an assigned percipient event, and the situation of the assemblage of causal components which is the situation of the 'real' object.

CHAPTER XVII

FIGURES

62. Sense-Figures. 62·1 There are two types of objects which can be included under the general name of 'figures'; objects of one type will be termed 'sense-figures,' and of the other type 'geometrical figures.'

Figures of either type arise from the perception of the relation of sense-objects to the properties which their situations have in respect to their relations of extension with other events. The primary type of figure is the sense-figure and the geometrical figure is derivative from it.

62·2 Every sort of sense-object will have its own peculiar sort of sense-figure. The sense-figures associated with some sorts of sense-objects (e.g. smells and tastes) are barely perceptible, whereas the sense-figures associated with other sorts of sense-objects (e.g. sights and touches) are of insistent obviousness. The condition that a sense-object should have a figure within a given duration can be precisely stated: A sense-object O, as perceived in a situation σ which extends throughout a duration d of a time-system α, possesses a figure in d, if every volume of σ, lying in a moment of α inherent in d, is congruent with every other such volume.

Owing to the inexactitude of perception small quantitative defects from the rigorous fulfilment of this condition do not in practice hinder the perception of figure. Namely, the possession of sense-figure follows from the sufficiently approximate fulfilment of this condition. The durations which are important from the point of view of sense-figures are those which form present durations of perceptions—in general, those durations which are cogredient with a percipient event and are each short enough to form one immediate present. Thus in the numerous instances in which there is no large change within such an immediate present, there is a perceived figure. Accordingly we can define a sense-figure precisely as follows:

The figure, for a time-system α, of sense-object O in situation σ is the relation holding, and only holding, between O and any α-volume congruent to a member of the set of α-volumes of σ.

This definition is only important when the α-volumes of σ are all nearly congruent to each other; because only in that case is this relation recognisable in perception.

62·3 Thus, each sense-object is primarily capable of its own sort of sense-figure and of that sort only. There are the sense-figures of blue of one shade, and the sense-figures of blue of another shade, and the separate sets of figures belonging to all the shades of reds and greens and yellows. There is the set of figures of the touch of velvet, and the set of figures of the touch of marble at particular temperatures of hand and surface and with a particular polish of surface.

62·4 But there is an analogy of sense-objects and this begets an analogy of figures. For example, there is an analogy between blues of all shades, and a corresponding analogy between their sets of figures. Each such analogy amid sense-objects issues in an object of a type not hitherto named. Call it the type of 'generalised sense-objects.' For example, we can recognise blue and ignore its particular shade. Correspondingly we can recognise a blue sense-figure, and ignore the differences between a light-blue sense-figure and a dark-blue sense-figure. We can go further, and recognise colour and ignore the particular colour; and correspondingly there are recognisable sight-figures underlying figures of particular shades of particular colours.

62·5 But it would be a mistake to insist on the derivation of the generalised sense-figures from the recognition of generalised sense-objects. In general the converse process would seem to be nearer the truth. Namely, the analogy amid sense-figures is more insistently perceptible than the analogy amid sense-objects; and the derivation is as much from the generalised sense-figure to the generalised sense-object as in the converse order.

We must go further than this. Perceptive insistency is not ranged in the order of simplicity as determined by a reflective analysis of the elements of our awareness of nature. Sense-figures possess a higher perceptive insistency than the corresponding sense-objects. We first notice a dark-blue figure and pass to the dark-blueness.

62·6 Indeed the high perceptive power of figures is at once the foundation of our natural knowledge and the origin of our philosophical errors. It has led the theory of space to be annexed to objects and not to events, and thus created the fatal divorce between space and time. A figure, being an object, is not in space or time, except in a derivative sense.

This perceptive power of figures carries us to the direct recognition of sorts of objects which otherwise would remain in the region of abstract logical concepts. For example, our perception of sight-figures leads to the recognition of colour as being what is common to all particular colours.

63. Geometrical Figures. 63·1 The generalisation
which introduces geometrical figures is an extreme instance of the sort of generalisation already considered.
Namely, instead of generalising from a dark-blue figure
to a sight-figure, we pass to the concept of the relation
of any sense-object to the volumes of its situation.
This concept of a figure, in which any particular sense
has been lost sight of, would be entirely without any
counterpart in perception, if it were not for the fact of
perceptual objects. A perceptual object is the association
in one situation of a set of sense-objects, in general
'conveyed' by the normal perception of one of them.
The high perceptive capacity of sense-figures leads to
their association in a generalised figure, which is the
geometrical figure of the object. Indeed, the insistent
obviousness of the geometrical figure is one reason for
the perception of perceptual objects. The object is not
the figure, but our awareness of it is derived from our
awareness of the figure. The reason for discriminating
the perceptual object from its figure in that situation
is that the physical object persists while its figure
changes. For example, a sock can be twisted into all
sorts of figures.

63·2 The current doctrine of different kinds of space
—tactual space, visual space, and so on—arises entirely
from the error of deducing space from the relations
between figures. With such a procedure, since there
are different types of figures for different types of sense,
evidently there must be different types of space for
different types of sense. And the demand created the
supply.

63·3 If however the modern assimilation of space
and time is to hold, we must then go further and admit

different kinds of time for different kinds of sense, namely a tactual time, a visual time, and so on. If this be allowed, it is difficult to understand how the disjecta membra of our perceptual experience manage to collect themselves into a common world.

For example, it would require a pre-established harmony to secure that the visual newspaper was delivered at the visual time of the visual breakfast in the visual room and also the tactual newspaper was delivered at the tactual time of the tactual breakfast in the tactual room. It is difficult enough for the plain man—such as the present author—to accept the miracle of getting the two newspapers into the two rooms daily with such admirable exactitude at the same time. But the additional miracle introduced by the two times is really incredible.

The procedure of this enquiry admits the different types of figures, but rejects the different types of space.

CHAPTER XVIII

RHYTHMS

64. Rhythms. 64·1 The theory of percipient objects is beyond the scope of this work of which the aim is to illustrate the principles of natural knowledge by an examination of the data and experiential laws fundamental for physical science. A percipient object is in some sense beyond nature.

But nature includes life; and the way of conceiving nature developed in the preceding chapter has its bearing on biological conceptions as to the sense in which life can be said to be thus included.

64·2 An object is a characteristic of an event. Such an object may be in fact a multiple relation between objects situated in various parts of the whole event. In this case the quality of the whole is the relationship between its parts, and the relation between the parts is the quality of the whole. The whole event being what it is, its parts have thereby certain defined relations; and the parts having all the relations which they do have, it follows that the whole event is what it is. The whole is explained by a full knowledge of the parts as situations of objects, and the parts by a full knowledge of the whole. Such an object is a pattern.

64·3 The discussion of life in nature has become canalised along certain conventional lines based upon the traditional concepts of science. We are aware of living objects. But the phrase 'living objects' is misleading; we should more accurately say, 'objects expressing life,' or 'life-bearing objects.' Namely, the individual life *is*, beyond the mere object. There is not an object which, after being known as an object, is then in itself judged to be alive. The specific recognisable liveliness is the recognised character of the relation of the object to the event which is its situation. Thus to say that the object is alive suppresses the necessary reference to the event; and to say that an event is alive suppresses the necessary reference to the object.

64·4 We have therefore to ask, what sort of events have life in their relations to objects situated in them, and what sort of objects have life in their relations to their situations? A life-bearing object is not an 'uniform' object. Life (as known to us) involves the completion of rhythmic parts within the life-bearing event which exhibits that object. We can diminish the time-parts, and, if the rhythms be unbroken, still discover the same object of life in the curtailed event. But if the diminution of the duration be carried to the extent of breaking the rhythm, the life-bearing object is no longer to be found as a quality of the slice of the original event cut off within that duration. This is no special peculiarity of life. It is equally true of a molecule of iron or of a musical phrase. Thus there is no such thing as life 'at one instant'; life is too obstinately concrete to be located in an extensive element of an instantaneous space.

64·5 The events which are associated by us with life are also the situations of physical objects. But the physical object though essential is not an adequate condition for its occurrence. A change in the object almost imperceptible from the physical point of view destroys the life in the succeeding situations of the object. The physical object, as apparent, is a material object and as such is uniform; but when we turn to the causal components of such an object, the apparent character of the whole situation is thereby superseded by the rhythmic quasi-periodic characters of a multitude of parts which are the situations of molecules.

In an analogous way we seek for a causal character of the event which in some way or another is apparent to us as alive, and we seek for an expression of this causal character in terms of the causal components of the physical object. It would seem therefore (if the analogy is to be pursued) that apparent life in any situation has, as its counterpart in that situation, more complex, subtler rhythms than those whose aggregate is essential for the physical object.

64·6 Furthermore in the physical object we have in a sense lost the rhythms in the macroscopic aggregate which is the final causal character. But life preserves its expression of rhythm and its sensitiveness to rhythm. Life is the rhythm as such, whereas a physical object is an average of rhythms which build no rhythm in their aggregation; and thus matter is in itself lifeless.

Life is complex in its expression, involving more than percipience, namely desire, emotion, will, and feeling. It exhibits variations of grade, higher and lower, such that the higher grade presupposes the lower for its very existence. This suggests a closer identification of rhythm as the causal counterpart of life; namely, that wherever there is some rhythm, there is some life, only perceptible to us when the analogies are sufficiently close. The rhythm is then the life, in the sense in which it can be said to be included within nature.

64·7 Now a rhythm is recognisable and is so far an object. But it is more than an object; for it is an object formed of other objects interwoven upon the background of essential change. A rhythm involves a pattern and to that extent is always self-identical. But no rhythm can be a mere pattern; for the rhythmic quality depends equally upon the differences involved in each exhibition of the pattern. The essence of rhythm is the fusion of sameness and novelty; so that the whole never loses the essential unity of the pattern, while the parts exhibit the contrast arising from the novelty of their detail. A mere recurrence kills rhythm as surely as does a mere confusion of differences. A crystal lacks rhythm from excess of pattern, while a fog is unrhythmic in that it exhibits a patternless confusion of detail. Again there are gradations of rhythm. The more perfect rhythm is built upon component rhythms. A subordinate part with crystalline excess of pattern or with foggy confusion weakens the rhythm. Thus every great rhythm presupposes lesser rhythms without which it could not be. No rhythm can be founded upon mere confusion or mere sameness.

64·8 An event, considered as gaining its unity from the continuity of extension and its unique novelty from its inherent character of 'passage,' contributes one factor to life; and the pattern exhibited within the event, which as self-identical should be a rigid recurrence, contributes the other factor to life. A rhythm is too concrete to be truly an object. It refuses to be disengaged from the event in the form of a true object which would be mere pattern. What the pattern does do is to impress its atomic character on a certain whole event which, as one whole bearing its atomic pattern, is a unique type of natural element, neither a mere event nor a mere object as object is here defined. This atomic character does not imply a discontinuous existence for a rhythm; thus a wave-length as marked out in various positions along a train of waves exhibits the whole rhythm of the train at each position of its continuous travel.

64·9 The very fact of a non-uniform object involves some rhythm. Such objects appear to our apprehension in events at certain stages of extensive size, provided that we confine attention to those organisms with stability of existence, each in close association with one physical object or with one set of causal material objects. Molecules are non-uniform objects and as such exhibit a rhythm; although, as known to us, it is a rhythm of excessive simplicity. Living bodies exhibit rhythm of the greatest subtlety within our apprehension. Solar systems and star clusters exhibit rhythm of a simplicity analogous to that of molecules. It is impossible not to suspect that the gain in apparent complexity at the stage of our own rhythm-bearing events is due rather to our angle of vision than to any inherent fact of nature.

There are also stray rhythms which pass over the face of nature utilising physical objects as mere transient vehicles for their expression. To some extent this is the case in living bodies, which exhibit a continual assimilation and rejection of material. But the subtlety of rhythm appears to require a certain stability of material.

64·91 Thus the permanence of the individual rhythm within nature is not absolutely associated with one definite set of material objects. But the connection for subtler rhythms is very close. So far as direct observation is concerned all that we know of the essential relations of life in nature is stated in two short poetic phrases. The obvious aspect by Tennyson,

> "Blow, bugle, blow, set the wild echoes flying,
> And answer, echoes, answer, dying, dying, dying."

Namely, Bergson's élan vital and its relapse into matter.

And Wordsworth with more depth,

> "The music in my heart I bore,
> Long after it was heard no more."

NOTES

Note I. The whole of Part II, i.e. Chapters V to VII, suffers from a vagueness of expression due to the fact that the implications of my ideas had not shaped themselves with sufficient emphasis in my mind. In the first place every entity is an abstraction and presupposes certain systematic types of relatedness to other things. There is no such thing as an entity which could be real on its own, though it happens to be related to other things. Again the development of these chapters presupposes that philosophy, even modern philosophy, has been unduly influenced by the Aristotelian categories, in particular those of substance, quantity, quality, despite the criticism to which these categories have been subjected. A detailed analysis of the complex notions which are concealed in the terms quantity and quality is required, but it cannot be given in a note. It was for this reason that I avoided terms such as 'Universal' which presuppose an outlook which is here repudiated. But in many respects the statement that an object is a universal does explain what I mean. Particularity attaches to events and to historical routes among events. But there is a flux of things transcending that of nature —in the narrow sense in which nature is here construed. Accordingly particularity cannot be confined to natural events. The whole subject requires fuller and more systematic treatment which I hope in the immediate future to undertake. The main point hinges onto the ingression* of objects into social entities, and onto the

* Cf. my *Concept of Nature* for a short introduction to the meaning of this term.

analysis of the process of the realisation of social entities.

In the list of objects in § 13·2 (p. 60), I was distinguishing the percipient object, which roughly speaking is an individual, as mental, from a percipient event which is the flux of experience of a living organism. But the percipient object is shadowy in this book and is clearly outside 'nature.'

Note II. The book is dominated by the idea [cf. § 14·1, p. 61] that the relation of extension has a unique preeminence and that everything can be got out of it. During the development of the theme, it gradually became evident that this is not the case, and cogredience [cf. § 16·4] had to be introduced. But the true doctrine, that 'process' is the fundamental idea, was not in my mind with sufficient emphasis. Extension is derivative from process, and is required by it.

This failure to insist properly on 'process' is the reason for the paradoxical air attaching to the statement that 'objects are only derivatively in space and time by reason of their relations to events' [cf. § 15·2]. Objects are of course essential for process, as appears clearly enough in the course of any analysis of process. But it is evident that particular times cannot result from the mere relations between objects which are at all times; and analogously for space. Accordingly space and time must result from something in process which transcends objects.

But natural objects require space and time, so that space and time belong to their relational essence without which they cannot be themselves.

In § 15·4 it is pointed out that it is viâ objects that the concept of possibility has application. This suggestion

requires further elaboration which cannot be attempted
here. Similarly in § 15·8, it is pointed out that continuity is derived from events, and atomicity from
objects. This also requires development. It must suffice
for the moment to suggest that a scientific object is an
atomic structure imposed upon the continuity of events.

Part II should be read in connection with Part IV at
the end of the book.

Note III. Chapter VI is made clearer by noting that
the present duration [cf. § 16·2] is primarily marked out
by the significance of an interconnected display of sensa
and of other associated objects immediately apparent.
The duration is the realisation of a social entity in which
the sense-objects and perceptual objects [cf. § 23·9] are
ingredient.

The antecedent physical objects [cf. § 24·5] and
scientific objects [cf. § 25] which occasion the duration
to be what it is are another story, and the persistent
habit of muddling the two sets of entities in philosophy
—following the lead of language—is the origin of much
confusion. For example there are four distinct meanings
according to which you can speak of a chair; you may
mean (i) a collection of sense-data, or (ii) a perceptual
object, or (iii) a physical object, or (iv) a collection of
scientific objects, such as molecules or electrons.

It will be noted that I now make a distinction between perceptual objects and physical objects, contrary
to § 24·5. Thus a physical object is a social entity resulting from scientific objects, and halfway towards a
perceptual object.

Physical objects and scientific objects are causal
characters which are discussed in Chapter XVI. Also

sense-objects and perceptual objects are the apparent characters discussed in that chapter. But the separation between the apparent and the causal must not be over stressed: it is relative to a deliberately limited point of view.

The whole of § 23 would be made clearer by the use of the term 'ingression' for the complex relation of a sense-object to the other factors of nature which make up a social entity in process of realisation. The over-simplification involved in the Aristotelian concept of 'quality-subject' has obscured the analysis of ingression.

Also § 24 is confused by a wavering between the 'class-theory' of perceptual objects and the 'control-theory' of physical objects, and by the confusion between perceptual and physical objects. I do not hold the class-theory now in any form, and was endeavouring in this book to get away from it. In Chapter XVI the 'causal character' is identical with the perceptual object so far as immediate perception is concerned, with the physical object so far as further discrimination of the significance of immediate appearance is concerned, and with scientific objects so far as more detailed analysis is concerned.

Note IV. The attempt in § 33 to define a duration merely by means of its unlimitedness is a failure. In a note to the *Concept of Nature*, I point out that there is an analogous unlimitedness through time, corresponding to the spatial unlimitedness of a duration. Both unlimitednesses arise from the uniform significance of spatial extension for one, and of temporal extension for the other. Thus there is a temporal unlimitedness arising from a stationary event [cf. § 41] in a definite space-time system. A point-track is the outcome of such

an unlimitedness deprived of all spatial extension. But
of course a point-track is a mere limit arrived at by the
method of extensive abstraction. But I still hold to
§ 35·4·

Note V. In § 47 the concept of normality is explained
with unnecessary elaboration. Let M_α and M_β be two
intersecting moments belonging to diverse space-time
systems α and β. Thus intersection of M_α and M_β is an
instantaneous plane $\pi_{\alpha\beta}$ which lies in the instantaneous

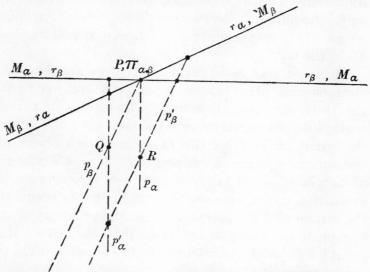

three-dimensional moments M_α and M_β. Let P be an
event-particle [i.e. an instantaneous point] in $\pi_{\alpha\beta}$. Then
there is **a point-track** p_α through P, belonging to the
space-time system α, and an analogous point-track
p_β belonging to β.

This conception is diagrammatically illustrated [but
not exactly represented] by the annexed diagram in
which two dimensions have been lost. Thus the three-

dimensional regions M_a and M_β are represented by two straight lines. These lines also represent [with full dimensions] two rects r_β and r_a in M_a and M_β respectively: the genesis of these rects will be explained below. The two-dimensional 'level' $\pi_{a\beta}$ is represented by the single event-particle P which lies in it: p_a is the a-point-track through P, and p_β the β-point-track through P; these are represented by dotted lines.

The plane of the paper is the two-dimensional matrix containing all the a-point-tracks through event-particles on p_β, and all the β-point-tracks through event-particles on p_a. This matrix intersects M_a in the rect r_β, and M_β in the rect r_a.

Then the rect r_β is perpendicular to the level, or instantaneous plane $\pi_{a\beta}$, where both r_β and $\pi_{a\beta}$ lie in the three-dimensional instantaneous space M_a; also analogously for the rect r_a and the level $\pi_{a\beta}$ in M_β. Thus the system of levels parallel to $\pi_{a\beta}$ in M_a is a system of instantaneous planes in M_a perpendicular to the system of rects in M_a parallel to r_β. This pair of systems reflects in M_a the relation between the system of β-moments to the system of β-point-tracks. Thus the geometry of an instantaneous moment expresses the relations of the event-particles of the moment to the whole bundle of alternative time-systems.

In this connection it may be well to expand the substance of a paragraph [pp. 89, 90] in the *Concept of Nature*:—An instantaneous point is an event-particle. It has two aspects. In one aspect it is there, where it is, in relation to the moments of the various space-time systems of the whole bundle of such systems. This aspect is expressed by the definition of a punct, which is determined by the individual moments (one from each space-time system) which contain it. The indivisibility of a point is expressed by the fact that any moment either contains the whole punct or contains nothing of the punct: the three-dimensionality of space, with time as a fourth dimension, is expressed by the fact that a punct is defined by four moments [not exceptionally related]: the position of the punct is expressed by those moments which do contain it. In another aspect a point is got at as a limit by indefinitely diminishing the dimensions of circumambient space-time. This aspect of absence of dimensions is expressed by the definition of an event-particle by means of abstractive sets with a certain quality of primeness in relation to puncts.

The intimate connection of geometry with the properties of the bundle of space-time systems is thus illustrated by the characters of instantaneous points, planes, and lines, and by the origin of parallelism and normality. Geometry expresses in three-dimensions these qualities of the four-dimensional space-time continuum.

Note VI. The deductions in §§ 50·3 and 51·1 are over condensed. They can be expanded as follows:

Substituting from (ii) and (iii) of § 50·3 in (iii) of 49·7, we find

$$1 = -\frac{\Omega_{\alpha\beta}'}{\Omega_{\beta\alpha}'} = -\frac{\Omega_{\alpha\beta}'''}{\Omega_{\beta\alpha}'''} = \frac{1}{\Omega_{\beta\alpha}{}^2 - \Omega_{\beta\alpha}'\Omega_{\beta\alpha}'''}$$
$$= \Omega_{\alpha\beta}{}^2 - \Omega_{\alpha\beta}'\Omega_{\alpha\beta}'''.$$

Thus
$$\Omega_{\alpha\beta}' + \Omega_{\beta\alpha}' = 0,$$
$$\Omega_{\alpha\beta}''' + \Omega_{\beta\alpha}''' = 0,$$
$$\Omega_{\alpha\beta}{}^2 - \Omega_{\alpha\beta}'\Omega_{\alpha\beta}''' = 1.$$

Hence by (iii) of 50·1
$$\Omega_{\alpha\beta}{}^2 + V_{\alpha\beta}\Omega_{\alpha\beta}\Omega_{\alpha\beta}''' = 1.$$

A CATALOGUE OF
SELECTED DOVER BOOKS
IN ALL FIELDS OF INTEREST

A CATALOGUE OF SELECTED DOVER
BOOKS IN ALL FIELDS OF INTEREST

RACKHAM'S COLOR ILLUSTRATIONS FOR WAGNER'S RING. Rackham's finest mature work—all 64 full-color watercolors in a faithful and lush interpretation of the *Ring*. Full-sized plates on coated stock of the paintings used by opera companies for authentic staging of Wagner. Captions aid in following complete Ring cycle. Introduction. 64 illustrations plus vignettes. 72pp. 8⅝ x 11¼. 23779-6 Pa. $6.00

CONTEMPORARY POLISH POSTERS IN FULL COLOR, edited by Joseph Czestochowski. 46 full-color examples of brilliant school of Polish graphic design, selected from world's first museum (near Warsaw) dedicated to poster art. Posters on circuses, films, plays, concerts all show cosmopolitan influences, free imagination. Introduction. 48pp. 9⅜ x 12¼. 23780-X Pa. $6.00

GRAPHIC WORKS OF EDVARD MUNCH, Edvard Munch. 90 haunting, evocative prints by first major Expressionist artist and one of the greatest graphic artists of his time: *The Scream, Anxiety, Death Chamber, The Kiss, Madonna,* etc. Introduction by Alfred Werner. 90pp. 9 x 12. 23765-6 Pa. $5.00

THE GOLDEN AGE OF THE POSTER, Hayward and Blanche Cirker. 70 extraordinary posters in full colors, from Maitres de l'Affiche, Mucha, Lautrec, Bradley, Cheret, Beardsley, many others. Total of 78pp. 9⅜ x 12¼. 22753-7 Pa. $5.95

THE NOTEBOOKS OF LEONARDO DA VINCI, edited by J. P. Richter. Extracts from manuscripts reveal great genius; on painting, sculpture, anatomy, sciences, geography, etc. Both Italian and English. 186 ms. pages reproduced, plus 500 additional drawings, including studies for *Last Supper*, Sforza monument, etc. 860pp. 7⅞ x 10¾. (Available in U.S. only) 22572-0, 22573-9 Pa., Two-vol. set $15.90

THE CODEX NUTTALL, as first edited by Zelia Nuttall. Only inexpensive edition, in full color, of a pre-Columbian Mexican (Mixtec) book. 88 color plates show kings, gods, heroes, temples, sacrifices. New explanatory, historical introduction by Arthur G. Miller. 96pp. 11⅜ x 8½. (Available in U.S. only) 23168-2 Pa. $7.95

UNE SEMAINE DE BONTÉ, A SURREALISTIC NOVEL IN COLLAGE, Max Ernst. Masterpiece created out of 19th-century periodical illustrations, explores worlds of terror and surprise. Some consider this Ernst's greatest work. 208pp. 8⅛ x 11. 23252-2 Pa. $6.00

DRAWINGS OF WILLIAM BLAKE, William Blake. 92 plates from Book of Job, *Divine Comedy, Paradise Lost,* visionary heads, mythological figures, Laocoon, etc. Selection, introduction, commentary by Sir Geoffrey Keynes. 178pp. 8⅛ x 11. 22303-5 Pa. $4.00

ENGRAVINGS OF HOGARTH, William Hogarth. 101 of Hogarth's greatest works: *Rake's Progress, Harlot's Progress, Illustrations for Hudibras, Before and After, Beer Street and Gin Lane,* many more. Full commentary. 256pp. 11 x 13¾. 22479-1 Pa. $12.95

DAUMIER: 120 GREAT LITHOGRAPHS, Honore Daumier. Wide-ranging collection of lithographs by the greatest caricaturist of the 19th century. Concentrates on eternally popular series on lawyers, on married life, on liberated women, etc. Selection, introduction, and notes on plates by Charles F. Ramus. Total of 158pp. 9⅜ x 12¼. 23512-2 Pa. $6.00

DRAWINGS OF MUCHA, Alphonse Maria Mucha. Work reveals drafts-man of highest caliber: studies for famous posters and paintings, render-ings for book illustrations and ads, etc. 70 works, 9 in color; including 6 items not drawings. Introduction. List of illustrations. 72pp. 9⅜ x 12¼. (Available in U.S. only) 23672-2 Pa. $4.00

GIOVANNI BATTISTA PIRANESI: DRAWINGS IN THE PIERPONT MORGAN LIBRARY, Giovanni Battista Piranesi. For first time ever all of Morgan Library's collection, world's largest. 167 illustrations of rare Piranesi drawings—archeological, architectural, decorative and visionary. Essay, detailed list of drawings, chronology, captions. Edited by Felice Stampfle. 144pp. 9⅜ x 12¼. 23714-1 Pa. $7.50

NEW YORK ETCHINGS (1905-1949), John Sloan. All of important American artist's N.Y. life etchings. 67 works include some of his best art; also lively historical record—Greenwich Village, tenement scenes. Edited by Sloan's widow. Introduction and captions. 79pp. 8⅜ x 11¼. 23651-X Pa. $4.00

CHINESE PAINTING AND CALLIGRAPHY: A PICTORIAL SURVEY, Wan-go Weng. 69 fine examples from John M. Crawford's matchless private collection: landscapes, birds, flowers, human figures, etc., plus calligraphy. Every basic form included: hanging scrolls, handscrolls, album leaves, fans, etc. 109 illustrations. Introduction. Captions. 192pp. 8⅞ x 11¾. 23707-9 Pa. $7.95

DRAWINGS OF REMBRANDT, edited by Seymour Slive. Updated Lipp-mann, Hofstede de Groot edition, with definitive scholarly apparatus. All portraits, biblical sketches, landscapes, nudes, Oriental figures, classical studies, together with selection of work by followers. 550 illustrations. Total of 630pp. 9⅛ x 12¼. 21485-0, 21486-9 Pa., Two-vol. set $15.00

THE DISASTERS OF WAR, Francisco Goya. 83 etchings record horrors of Napoleonic wars in Spain and war in general. Reprint of 1st edition, plus 3 additional plates. Introduction by Philip Hofer. 97pp. 9⅜ x 8¼. 21872-4 Pa. $4.00

CATALOGUE OF DOVER BOOKS

THE EARLY WORK OF AUBREY BEARDSLEY, Aubrey Beardsley. 157 plates, 2 in color: *Manon Lescaut, Madame Bovary, Morte Darthur, Salome,* other. Introduction by H. Marillier. 182pp. 8⅛ x 11. 21816-3 Pa. $4.50

THE LATER WORK OF AUBREY BEARDSLEY, Aubrey Beardsley. Exotic masterpieces of full maturity: *Venus and Tannhauser, Lysistrata, Rape of the Lock, Volpone,* Savoy material, etc. 174 plates, 2 in color. 186pp. 8⅛ x 11. 21817-1 Pa. $5.95

THOMAS NAST'S CHRISTMAS DRAWINGS, Thomas Nast. Almost all Christmas drawings by creator of image of Santa Claus as we know it, and one of America's foremost illustrators and political cartoonists. 66 illustrations. 3 illustrations in color on covers. 96pp. 8⅜ x 11¼. 23660-9 Pa. $3.50

THE DORÉ ILLUSTRATIONS FOR DANTE'S DIVINE COMEDY, Gustave Doré. All 135 plates from Inferno, Purgatory, Paradise; fantastic tortures, infernal landscapes, celestial wonders. Each plate with appropriate (translated) verses. 141pp. 9 x 12. 23231-X Pa. $4.50

DORÉ'S ILLUSTRATIONS FOR RABELAIS, Gustave Doré. 252 striking illustrations of *Gargantua and Pantagruel* books by foremost 19th-century illustrator. Including 60 plates, 192 delightful smaller illustrations. 153pp. 9 x 12. 23656-0 Pa. $5.00

LONDON: A PILGRIMAGE, Gustave Doré, Blanchard Jerrold. Squalor, riches, misery, beauty of mid-Victorian metropolis; 55 wonderful plates, 125 other illustrations, full social, cultural text by Jerrold. 191pp. of text. 9⅜ x 12¼. 22306-X Pa. $7.00

THE RIME OF THE ANCIENT MARINER, Gustave Doré, S. T. Coleridge. Dore's finest work, 34 plates capture moods, subtleties of poem. Full text. Introduction by Millicent Rose. 77pp. 9¼ x 12. 22305-1 Pa. $3.50

THE DORE BIBLE ILLUSTRATIONS, Gustave Doré. All wonderful, detailed plates: Adam and Eve, Flood, Babylon, Life of Jesus, etc. Brief King James text with each plate. Introduction by Millicent Rose. 241 plates. 241pp. 9 x 12. 23004-X Pa. $6.00

THE COMPLETE ENGRAVINGS, ETCHINGS AND DRYPOINTS OF ALBRECHT DURER. "Knight, Death and Devil"; "Melencolia," and more—all Dürer's known works in all three media, including 6 works formerly attributed to him. 120 plates. 235pp. 8⅜ x 11¼. 22851-7 Pa. $6.50

MECHANICK EXERCISES ON THE WHOLE ART OF PRINTING, Joseph Moxon. First complete book (1683-4) ever written about typography, a compendium of everything known about printing at the latter part of 17th century. Reprint of 2nd (1962) Oxford Univ. Press edition. 74 illustrations. Total of 550pp. 6⅛ x 9¼. 23617-X Pa. $7.95

THE COMPLETE WOODCUTS OF ALBRECHT DURER, edited by
Dr. W. Kurth. 346 in all: "Old Testament," "St. Jerome," "Passion,"
"Life of Virgin," Apocalypse," many others. Introduction by Campbell
Dodgson. 285pp. 8½ x 12¼.   21097-9 Pa. $7.50

DRAWINGS OF ALBRECHT DURER, edited by Heinrich Wölfflin. 81
plates show development from youth to full style. Many favorites; many
new. Introduction by Alfred Werner. 96pp. 8⅛ x 11.   22352-3 Pa. $5.00

THE HUMAN FIGURE, Albrecht Dürer. Experiments in various tech-
niques—stereometric, progressive proportional, and others. Also life studies
that rank among finest ever done. Complete reprinting of Dresden Sketch-
book. 170 plates. 355pp. 8⅜ x 11¼.   21042-1 Pa. $7.95

OF THE JUST SHAPING OF LETTERS, Albrecht Dürer. Renaissance
artist explains design of Roman majuscules by geometry, also Gothic lower
and capitals. Grolier Club edition. 43pp. 7⅞ x 10¾   21306-4 Pa. $3.00

TEN BOOKS ON ARCHITECTURE, Vitruvius. The most important book
ever written on architecture. Early Roman aesthetics, technology, classical
orders, site selection, all other aspects. Stands behind everything since.
Morgan translation. 331pp. 5⅜ x 8½.   20645-9 Pa. $4.50

THE FOUR BOOKS OF ARCHITECTURE, Andrea Palladio. 16th-century
classic responsible for Palladian movement and style. Covers classical architectural remains, Renaissance revivals, classical orders, etc. 1738 Ware
English edition. Introduction by A. Placzek. 216 plates. 110pp. of text.
9½ x 12¾.   21308-0 Pa. $10.00

HORIZONS, Norman Bel Geddes. Great industrialist stage designer, "father
of streamlining," on application of aesthetics to transportation, amusement,
architecture, etc. 1932 prophetic account; function, theory, specific projects.
222 illustrations. 312pp. 7⅞ x 10¾.   23514-9 Pa. $6.95

FRANK LLOYD WRIGHT'S FALLINGWATER, Donald Hoffmann. Full,
illustrated story of conception and building of Wright's masterwork at
Bear Run, Pa. 100 photographs of site, construction, and details of completed structure. 112pp. 9¼ x 10.   23671-4 Pa. $5.50

THE ELEMENTS OF DRAWING, John Ruskin. Timeless classic by great
Vittorian; starts with basic ideas, works through more difficult. Many
practical exercises. 48 illustrations. Introduction by Lawrence Campbell.
228pp. 5⅜ x 8½.   22730-8 Pa. $3.75

GIST OF ART, John Sloan. Greatest modern American teacher, Art Students League, offers innumerable hints, instructions, guided comments to
help you in painting. Not a formal course. 46 illustrations. Introduction
by Helen Sloan. 200pp. 5⅜ x 8½.   23435-5 Pa. $4.00

CATALOGUE OF DOVER BOOKS

THE ANATOMY OF THE HORSE, George Stubbs. Often considered the great masterpiece of animal anatomy. Full reproduction of 1766 edition, plus prospectus; original text and modernized text. 36 plates. Introduction by Eleanor Garvey. 121pp. 11 x 14¾. 23402-9 Pa. $6.00

BRIDGMAN'S LIFE DRAWING, George B. Bridgman. More than 500 illustrative drawings and text teach you to abstract the body into its major masses, use light and shade, proportion; as well as specific areas of anatomy, of which Bridgman is master. 192pp. 6½ x 9¼. (Available in U.S. only) 22710-3 Pa. $3.50

ART NOUVEAU DESIGNS IN COLOR, Alphonse Mucha, Maurice Verneuil, Georges Auriol. Full-color reproduction of *Combinaisons ornementales* (c. 1900) by Art Nouveau masters. Floral, animal, geometric, interlacings, swashes—borders, frames, spots—all incredibly beautiful. 60 plates, hundreds of designs. 9⅜ x 8-1/16. 22885-1 Pa. $4.00

FULL-COLOR FLORAL DESIGNS IN THE ART NOUVEAU STYLE, E. A. Seguy. 166 motifs, on 40 plates, from *Les fleurs et leurs applications decoratives* (1902): borders, circular designs, repeats, allovers, "spots." All in authentic Art Nouveau colors. 48pp. 9⅜ x 12¼. 23439-8 Pa. $5.00

A DIDEROT PICTORIAL ENCYCLOPEDIA OF TRADES AND INDUSTRY, edited by Charles C. Gillispie. 485 most interesting plates from the great French Encyclopedia of the 18th century show hundreds of working figures, artifacts, process, land and cityscapes; glassmaking, papermaking, metal extraction, construction, weaving, making furniture, clothing, wigs, dozens of other activities. Plates fully explained. 920pp. 9 x 12. 22284-5, 22285-3 Clothbd., Two-vol. set $40.00

HANDBOOK OF EARLY ADVERTISING ART, Clarence P. Hornung. Largest collection of copyright-free early and antique advertising art ever compiled. Over 6,000 illustrations, from Franklin's time to the 1890's for special effects, novelty. Valuable source, almost inexhaustible.
Pictorial Volume. Agriculture, the zodiac, animals, autos, birds, Christmas, fire engines, flowers, trees, musical instruments, ships, games and sports, much more. Arranged by subject matter and use. 237 plates. 288pp. 9 x 12. 20122-8 Clothbd. $14.50

Typographical Volume. Roman and Gothic faces ranging from 10 point to 300 point, "Barnum," German and Old English faces, script, logotypes, scrolls and flourishes, 1115 ornamental initials, 67 complete alphabets, more. 310 plates. 320pp. 9 x 12. 20123-6 Clothbd. $15.00

CALLIGRAPHY (CALLIGRAPHIA LATINA), J. G. Schwandner. High point of 18th-century ornamental calligraphy. Very ornate initials, scrolls, borders, cherubs, birds, lettered examples. 172pp. 9 x 13. 20475-8 Pa. $7.00

CATALOGUE OF DOVER BOOKS

ART FORMS IN NATURE, Ernst Haeckel. Multitude of strangely beautiful natural forms: Radiolaria, Foraminifera, jellyfishes, fungi, turtles, bats, etc. All 100 plates of the 19th-century evolutionist's *Kunstformen der Natur* (1904). 100pp. 9⅜ x 12¼. 22987-4 Pa. $5.00

CHILDREN: A PICTORIAL ARCHIVE FROM NINETEENTH-CENTURY SOURCES, edited by Carol Belanger Grafton. 242 rare, copyright-free wood engravings for artists and designers. Widest such selection available. All illustrations in line. 119pp. 8⅜ x 11¼. 23694-3 Pa. $4.00

WOMEN: A PICTORIAL ARCHIVE FROM NINETEENTH-CENTURY SOURCES, edited by Jim Harter. 391 copyright-free wood engravings for artists and designers selected from rare periodicals. Most extensive such collection available. All illustrations in line. 128pp. 9 x 12. 23703-6 Pa. $4.50

ARABIC ART IN COLOR, Prisse d'Avennes. From the greatest ornamentalists of all time—50 plates in color, rarely seen outside the Near East, rich in suggestion and stimulus. Includes 4 plates on covers. 46pp. 9⅜ x 12¼. 23658-7 Pa. $6.00

AUTHENTIC ALGERIAN CARPET DESIGNS AND MOTIFS, edited by June Beveridge. Algerian carpets are world famous. Dozens of geometrical motifs are charted on grids, color-coded, for weavers, needleworkers, craftsmen, designers. 53 illustrations plus 4 in color. 48pp. 8¼ x 11. (Available in U.S. only) 23650-1 Pa. $1.75

DICTIONARY OF AMERICAN PORTRAITS, edited by Hayward and Blanche Cirker. 4000 important Americans, earliest times to 1905, mostly in clear line. Politicians, writers, soldiers, scientists, inventors, industrialists, Indians, Blacks, women, outlaws, etc. Identificatory information. 756pp. 9¼ x 12¾. 21823-6 Clothbd. $40.00

HOW THE OTHER HALF LIVES, Jacob A. Riis. Journalistic record of filth, degradation, upward drive in New York immigrant slums, shops, around 1900. New edition includes 100 original Riis photos, monuments of early photography. 233pp. 10 x 7⅞. 22012-5 Pa. $7.00

NEW YORK IN THE THIRTIES, Berenice Abbott. Noted photographer's fascinating study of city shows new buildings that have become famous and old sights that have disappeared forever. Insightful commentary. 97 photographs. 97pp. 11⅜ x 10. 22967-X Pa. $5.00

MEN AT WORK, Lewis W. Hine. Famous photographic studies of construction workers, railroad men, factory workers and coal miners. New supplement of 18 photos on Empire State building construction. New introduction by Jonathan L. Doherty. Total of 69 photos. 63pp. 8 x 10¾. 23475-4 Pa. $3.00

CATALOGUE OF DOVER BOOKS

THE DEPRESSION YEARS AS PHOTOGRAPHED BY ARTHUR ROTHSTEIN, Arthur Rothstein. First collection devoted entirely to the work of outstanding 1930s photographer: famous dust storm photo, ragged children, unemployed, etc. 120 photographs. Captions. 119pp. 9¼ x 10¾.
23590-4 Pa. $5.00

CAMERA WORK: A PICTORIAL GUIDE, Alfred Stieglitz. All 559 illustrations and plates from the most important periodical in the history of art photography, Camera Work (1903-17). Presented four to a page, reduced in size but still clear, in strict chronological order, with complete captions. Three indexes. Glossary. Bibliography. 176pp. 8⅜ x 11¼.
23591-2 Pa. $6.95

ALVIN LANGDON COBURN, PHOTOGRAPHER, Alvin L. Coburn. Revealing autobiography by one of greatest photographers of 20th century gives insider's version of Photo-Secession, plus comments on his own work. 77 photographs by Coburn. Edited by Helmut and Alison Gernsheim. 160pp. 8⅛ x 11.
23685-4 Pa. $6.00

NEW YORK IN THE FORTIES, Andreas Feininger. 162 brilliant photographs by the well-known photographer, formerly with Life magazine, show commuters, shoppers, Times Square at night, Harlem nightclub, Lower East Side, etc. Introduction and full captions by John von Hartz. 181pp. 9¼ x 10¾.
23585-8 Pa. $6.95

GREAT NEWS PHOTOS AND THE STORIES BEHIND THEM, John Faber. Dramatic volume of 140 great news photos, 1855 through 1976, and revealing stories behind them, with both historical and technical information. Hindenburg disaster, shooting of Oswald, nomination of Jimmy Carter, etc. 160pp. 8¼ x 11.
23667-6 Pa. $5.00

THE ART OF THE CINEMATOGRAPHER, Leonard Maltin. Survey of American cinematography history and anecdotal interviews with 5 masters—Arthur Miller, Hal Mohr, Hal Rosson, Lucien Ballard, and Conrad Hall. Very large selection of behind-the-scenes production photos. 105 photographs. Filmographies. Index. Originally Behind the Camera. 144pp. 8¼ x 11.
23686-2 Pa. $5.00

DESIGNS FOR THE THREE-CORNERED HAT (LE TRICORNE), Pablo Picasso. 32 fabulously rare drawings—including 31 color illustrations of costumes and accessories—for 1919 production of famous ballet. Edited by Parmenia Migel, who has written new introduction. 48pp. 9⅜ x 12¼. (Available in U.S. only)
23709-5 Pa. $5.00

NOTES OF A FILM DIRECTOR, Sergei Eisenstein. Greatest Russian filmmaker explains montage, making of Alexander Nevsky, aesthetics; comments on self, associates, great rivals (Chaplin), similar material. 78 illustrations. 240pp. 5⅜ x 8½.
22392-2 Pa. $4.50

CATALOGUE OF DOVER BOOKS

HOLLYWOOD GLAMOUR PORTRAITS, edited by John Kobal. 145 photos capture the stars from 1926-49, the high point in portrait photography. Gable, Harlow, Bogart, Bacall, Hedy Lamarr, Marlene Dietrich, Robert Montgomery, Marlon Brando, Veronica Lake; 94 stars in all. Full background on photographers, technical aspects, much more. Total of 160pp. 8⅜ x 11¼.   23352-9 Pa. $6.00

THE NEW YORK STAGE: FAMOUS PRODUCTIONS IN PHOTO-GRAPHS, edited by Stanley Appelbaum. 148 photographs from Museum of City of New York show 142 plays, 1883-1939. *Peter Pan, The Front Page, Dead End, Our Town,* O'Neill, hundreds of actors and actresses, etc. Full indexes. 154pp. 9½ x 10.   23241-7 Pa. $6.00

DIALOGUES CONCERNING TWO NEW SCIENCES, Galileo Galilei. Encompassing 30 years of experiment and thought, these dialogues deal with geometric demonstrations of fracture of solid bodies, cohesion, leverage, speed of light and sound, pendulums, falling bodies, accelerated motion, etc. 300pp. 5⅜ x 8½.   60099-8 Pa. $4.00

THE GREAT OPERA STARS IN HISTORIC PHOTOGRAPHS, edited by James Camner. 343 portraits from the 1850s to the 1940s: Tamburini, Mario, Caliapin, Jeritza, Melchior, Melba, Patti, Pinza, Schipa, Caruso, Farrar, Steber, Gobbi, and many more—270 performers in all. Index. 199pp. 8⅜ x 11¼.   23575-0 Pa. $7.50

J. S. BACH, Albert Schweitzer. Great full-length study of Bach, life, background to music, music, by foremost modern scholar. Ernest Newman translation. 650 musical examples. Total of 928pp. 5⅜ x 8½. (Available in U.S. only)   21631-4, 21632-2 Pa., Two-vol. set $11.00

COMPLETE PIANO SONATAS, Ludwig van Beethoven. All sonatas in the fine Schenker edition, with fingering, analytical material. One of best modern editions. Total of 615pp. 9 x 12. (Available in U.S. only)
          23134-8, 23135-6 Pa., Two-vol. set $15.50

KEYBOARD MUSIC, J. S. Bach. Bach-Gesellschaft edition. For harpsichord, piano, other keyboard instruments. English Suites, French Suites, Six Partitas, Goldberg Variations, Two-Part Inventions, Three-Part Sinfonias. 312pp. 8⅛ x 11. (Available in U.S. only)   22360-4 Pa. $6.95

FOUR SYMPHONIES IN FULL SCORE, Franz Schubert. Schubert's four most popular symphonies: No. 4 in C Minor ("Tragic"); No. 5 in B-flat Major; No. 8 in B Minor ("Unfinished"); No. 9 in C Major ("Great"). Breitkopf & Hartel edition. Study score. 261pp. 9⅜ x 12¼.
                    23681-1 Pa. $6.50

THE AUTHENTIC GILBERT & SULLIVAN SONGBOOK, W. S. Gilbert, A. S. Sullivan. Largest selection available; 92 songs, uncut, original keys, in piano rendering approved by Sullivan. Favorites and lesser-known fine numbers. Edited with plot synopses by James Spero. 3 illustrations. 399pp. 9 x 12.   23482-7 Pa. $9.95

PRINCIPLES OF ORCHESTRATION, Nikolay Rimsky-Korsakov. Great classical orchestrator provides fundamentals of tonal resonance, progression of parts, voice and orchestra, tutti effects, much else in major document. 330pp. of musical excerpts. 489pp. 6½ x 9¼. 21266-1 Pa. **$7.50**

TRISTAN UND ISOLDE, Richard Wagner. Full orchestral score with complete instrumentation. Do not confuse with piano reduction. Commentary by Felix Mottl, great Wagnerian conductor and scholar. Study score. 655pp. 8⅛ x 11. 22915-7 Pa. **$13.95**

REQUIEM IN FULL SCORE, Giuseppe Verdi. Immensely popular with choral groups and music lovers. Republication of edition published by C. F. Peters, Leipzig, n. d. German frontmaker in English translation. Glossary. Text in Latin. Study score. 204pp. 9⅜ x 12¼.
23682-X Pa. **$6.00**

COMPLETE CHAMBER MUSIC FOR STRINGS, Felix Mendelssohn. All of Mendelssohn's chamber music: Octet, 2 Quintets, 6 Quartets, and Four Pieces for String Quartet. (Nothing with piano is included). Complete works edition (1874-7). Study score. 283 pp. 9⅜ x 12¼.
23679-X Pa. **$7.50**

POPULAR SONGS OF NINETEENTH-CENTURY AMERICA, edited by Richard Jackson. 64 most important songs: "Old Oaken Bucket," "Arkansas Traveler," "Yellow Rose of Texas," etc. Authentic original sheet music, full introduction and commentaries. 290pp. 9 x 12. 23270-0 Pa. **$7.95**

COLLECTED PIANO WORKS, Scott Joplin. Edited by Vera Brodsky Lawrence. Practically all of Joplin's piano works—rags, two-steps, marches, waltzes, etc., 51 works in all. Extensive introduction by Rudi Blesh. Total of 345pp. 9 x 12. 23106-2 Pa. **$14.95**

BASIC PRINCIPLES OF CLASSICAL BALLET, Agrippina Vaganova. Great Russian theoretician, teacher explains methods for teaching classical ballet; incorporates best from French, Italian, Russian schools. 118 illustrations. 175pp. 5⅜ x 8½. 22036-2 Pa. **$2.50**

CHINESE CHARACTERS, L. Wieger. Rich analysis of 2300 characters according to traditional systems into primitives. Historical-semantic analysis to phonetics (Classical Mandarin) and radicals. 820pp. 6⅛ x 9¼.
21321-8 Pa. **$10.00**

EGYPTIAN LANGUAGE: EASY LESSONS IN EGYPTIAN HIERO-GLYPHICS, E. A. Wallis Budge. Foremost Egyptologist offers Egyptian grammar, explanation of hieroglyphics, many reading texts, dictionary of symbols. 246pp. 5 x 7½. (Available in U.S. only)
21394-3 Clothbd. **$7.50**

AN ETYMOLOGICAL DICTIONARY OF MODERN ENGLISH, Ernest Weekley. Richest, fullest work, by foremost British lexicographer. Detailed word histories. Inexhaustible. Do not confuse this with *Concise Etymological Dictionary,* which is abridged. Total of 856pp. 6½ x 9¼.
21873-2, 21874-0 Pa., Two-vol. set **$12.00**

CATALOGUE OF DOVER BOOKS

A MAYA GRAMMAR, Alfred M. Tozzer. Practical, useful English-language grammar by the Harvard anthropologist who was one of the three greatest American scholars in the area of Maya culture. Phonetics, grammatical processes, syntax, more. 301pp. 5⅜ x 8½.　　23465-7 Pa. $4.00

THE JOURNAL OF HENRY D. THOREAU, edited by Bradford Torrey, F. H. Allen. Complete reprinting of 14 volumes, 1837-61, over two million words; the sourcebooks for *Walden*, etc. Definitive. All original sketches, plus 75 photographs. Introduction by Walter Harding. Total of 1804pp. 8½ x 12¼.　　20312-3, 20313-1 Clothbd., Two-vol. set $70.00

CLASSIC GHOST STORIES, Charles Dickens and others. 18 wonderful stories you've wanted to reread: "The Monkey's Paw," "The House and the Brain," "The Upper Berth," "The Signalman," "Dracula's Guest," "The Tapestried Chamber," etc. Dickens, Scott, Mary Shelley, Stoker, etc. 330pp. 5⅜ x 8½.　　20735-8 Pa. $4.50

SEVEN SCIENCE FICTION NOVELS, H. G. Wells. Full novels. *First Men in the Moon, Island of Dr. Moreau, War of the Worlds, Food of the Gods, Invisible Man, Time Machine, In the Days of the Comet.* A basic science-fiction library. 1015pp. 5⅜ x 8½. (Available in U.S. only)
20264-X Clothbd. $8.95

ARMADALE, Wilkie Collins. Third great mystery novel by the author of *The Woman in White* and *The Moonstone.* Ingeniously plotted narrative shows an exceptional command of character, incident and mood. Original magazine version with 40 illustrations. 597pp. 5⅜ x 8½.
23429-0 Pa. $6.00

MASTERS OF MYSTERY, H. Douglas Thomson. The first book in English (1931) devoted to history and aesthetics of detective story. Poe, Doyle, LeFanu, Dickens, many others, up to 1930. New introduction and notes by E. F. Bleiler. 288pp. 5⅜ x 8½. (Available in U.S. only)
23606-4 Pa. $4.00

FLATLAND, E. A. Abbott. Science-fiction classic explores life of 2-D being in 3-D world. Read also as introduction to thought about hyperspace. Introduction by Banesh Hoffmann. 16 illustrations. 103pp. 5⅜ x 8½.
20001-9 Pa. $2.00

THREE SUPERNATURAL NOVELS OF THE VICTORIAN PERIOD, edited, with an introduction, by E. F. Bleiler. Reprinted complete and unabridged, three great classics of the supernatural: *The Haunted Hotel* by Wilkie Collins, *The Haunted House at Latchford* by Mrs. J. H. Riddell, and *The Lost Stradivarius* by J. Meade Falkner. 325pp. 5⅜ x 8½.
22571-2 Pa. $4.00

AYESHA: THE RETURN OF "SHE," H. Rider Haggard. Virtuoso sequel featuring the great mythic creation, Ayesha, in an adventure that is fully as good as the first book, *She.* Original magazine version, with 47 original illustrations by Maurice Greiffenhagen. 189pp. 6½ x 9¼.
23649-8 Pa. $3.50

UNCLE SILAS, J. Sheridan LeFanu. Victorian Gothic mystery novel, considered by many best of period, even better than Collins or Dickens. Wonderful psychological terror. Introduction by Frederick Shroyer. 436pp. 5⅜ x 8½. 21715-9 Pa. $6.00

JURGEN, James Branch Cabell. The great erotic fantasy of the 1920's that delighted thousands, shocked thousands more. Full final text, Lane edition with 13 plates by Frank Pape. 346pp. 5⅜ x 8½. 23507-6 Pa. $4.50

THE CLAVERINGS, Anthony Trollope. Major novel, chronicling aspects of British Victorian society, personalities. Reprint of Cornhill serialization, 16 plates by M. Edwards; first reprint of full text. Introduction by Norman Donaldson. 412pp. 5⅜ x 8½. 23464-9 Pa. $5.00

KEPT IN THE DARK, Anthony Trollope. Unusual short novel about Victorian morality and abnormal psychology by the great English author. Probably the first American publication. Frontispiece by Sir John Millais. 92pp. 6½ x 9¼. 23609-9 Pa. $2.50

RALPH THE HEIR, Anthony Trollope. Forgotten tale of illegitimacy, inheritance. Master novel of Trollope's later years. Victorian country estates, clubs, Parliament, fox hunting, world of fully realized characters. Reprint of 1871 edition. 12 illustrations by F. A. Faser. 434pp. of text. 5⅜ x 8½. 23642-0 Pa. $5.00

YEKL and THE IMPORTED BRIDEGROOM AND OTHER STORIES OF THE NEW YORK GHETTO, Abraham Cahan. Film *Hester Street* based on *Yekl* (1896). Novel, other stories among first about Jewish immigrants of N.Y.'s East Side. Highly praised by W. D. Howells—Cahan "a new star of realism." New introduction by Bernard G. Richards. 240pp. 5⅜ x 8½. 22427-9 Pa. $3.50

THE HIGH PLACE, James Branch Cabell. Great fantasy writer's enchanting comedy of disenchantment set in 18th-century France. Considered by some critics to be even better than his famous *Jurgen*. 10 illustrations and numerous vignettes by noted fantasy artist Frank C. Pape. 320pp. 5⅜ x 8½. 23670-6 Pa. $4.00

ALICE'S ADVENTURES UNDER GROUND, Lewis Carroll. Facsimile of ms. Carroll gave Alice Liddell in 1864. Different in many ways from final Alice. Handlettered, illustrated by Carroll. Introduction by Martin Gardner. 128pp. 5⅜ x 8½. 21482-6 Pa. $2.50

FAVORITE ANDREW LANG FAIRY TALE BOOKS IN MANY COLORS, Andrew Lang. The four Lang favorites in a boxed set—the complete *Red, Green, Yellow* and *Blue* Fairy Books. 164 stories; 439 illustrations by Lancelot Speed, Henry Ford and G. P. Jacomb Hood. Total of about 1500pp. 5⅜ x 8½. 23407-X Boxed set, Pa. $15.95

HOUSEHOLD STORIES BY THE BROTHERS GRIMM. All the great Grimm stories: "Rumpelstiltskin," "Snow White," "Hansel and Gretel," etc., with 114 illustrations by Walter Crane. 269pp. 5⅜ x 8½.

21080-4 Pa. $3.50

SLEEPING BEAUTY, illustrated by Arthur Rackham. Perhaps the fullest, most delightful version ever, told by C. S. Evans. Rackham's best work. 49 illustrations. 110pp. 7⅞ x 10¾.

22756-1 Pa. $2.50

AMERICAN FAIRY TALES, L. Frank Baum. Young cowboy lassoes Father Time; dummy in Mr. Floman's department store window comes to life; and 10 other fairy tales. 41 illustrations by N. P. Hall, Harry Kennedy, Ike Morgan, and Ralph Gardner. 209pp. 5⅜ x 8½.

23643-9 Pa. $3.00

THE WONDERFUL WIZARD OF OZ, L. Frank Baum. Facsimile in full color of America's finest children's classic. Introduction by Martin Gardner. 143 illustrations by W. W. Denslow. 267pp. 5⅜ x 8½.

20691-2 Pa. $3.50

THE TALE OF PETER RABBIT, Beatrix Potter. The inimitable Peter's terrifying adventure in Mr. McGregor's garden, with all 27 wonderful, full-color Potter illustrations. 55pp. 4¼ x 5½. (Available in U.S. only)

22827-4 Pa. $1.25

THE STORY OF KING ARTHUR AND HIS KNIGHTS, Howard Pyle. Finest children's version of life of King Arthur. 48 illustrations by Pyle. 131pp. 6⅛ x 9¼.

21445-1 Pa. $4.95

CARUSO'S CARICATURES, Enrico Caruso. Great tenor's remarkable caricatures of self, fellow musicians, composers, others. Toscanini, Puccini, Farrar, etc. Impish, cutting, insightful. 473 illustrations. Preface by M. Sisca. 217pp. 8⅜ x 11¼.

23528-9 Pa. $6.95

PERSONAL NARRATIVE OF A PILGRIMAGE TO ALMADINAH AND MECCAH, Richard Burton. Great travel classic by remarkably colorful personality. Burton, disguised as a Moroccan, visited sacred shrines of Islam, narrowly escaping death. Wonderful observations of Islamic life, customs, personalities. 47 illustrations. Total of 959pp. 5⅜ x 8½.

21217-3, 21218-1 Pa., Two-vol. set $12.00

INCIDENTS OF TRAVEL IN YUCATAN, John L. Stephens. Classic (1843) exploration of jungles of Yucatan, looking for evidences of Maya civilization. Travel adventures, Mexican and Indian culture, etc. Total of 669pp. 5⅜ x 8½.

20926-1, 20927-X Pa., Two-vol. set $7.90

AMERICAN LITERARY AUTOGRAPHS FROM WASHINGTON IRVING TO HENRY JAMES, Herbert Cahoon, et al. Letters, poems, manuscripts of Hawthorne, Thoreau, Twain, Alcott, Whitman, 67 other prominent American authors. Reproductions, full transcripts and commentary. Plus checklist of all American Literary Autographs in The Pierpont Morgan Library. Printed on exceptionally high-quality paper. 136 illustrations. 212pp. 9⅛ x 12¼.

23548-3 Pa. $12.50

AN AUTOBIOGRAPHY, Margaret Sanger. Exciting personal account of hard-fought battle for woman's right to birth control, against prejudice, church, law. Foremost feminist document. 504pp. 5⅜ x 8½.
20470-7 Pa. $5.50

MY BONDAGE AND MY FREEDOM, Frederick Douglass. Born as a slave, Douglass became outspoken force in antislavery movement. The best of Douglass's autobiographies. Graphic description of slave life. Introduction by P. Foner. 464pp. 5⅜ x 8½.
22457-0 Pa. $5.50

LIVING MY LIFE, Emma Goldman. Candid, no holds barred account by foremost American anarchist: her own life, anarchist movement, famous contemporaries, ideas and their impact. Struggles and confrontations in America, plus deportation to U.S.S.R. Shocking inside account of persecution of anarchists under Lenin. 13 plates. Total of 944pp. 5⅜ x 8½.
22543-7, 22544-5 Pa., Two-vol. set $12.00

LETTERS AND NOTES ON THE MANNERS, CUSTOMS AND CONDITIONS OF THE NORTH AMERICAN INDIANS, George Catlin. Classic account of life among Plains Indians: ceremonies, hunt, warfare, etc. Dover edition reproduces for first time all original paintings. 312 plates. 572pp. of text. 6⅛ x 9¼.
22118-0, 22119-9 Pa.. Two-vol. set $12.00

THE MAYA AND THEIR NEIGHBORS, edited by Clarence L. Hay, others. Synoptic view of Maya civilization in broadest sense, together with Northern, Southern neighbors. Integrates much background, valuable detail not elsewhere. Prepared by greatest scholars: Kroeber, Morley, Thompson, Spinden, Vaillant, many others. Sometimes called Tozzer Memorial Volume. 60 illustrations, linguistic map. 634pp. 5⅜ x 8½.
23510-6 Pa. $10.00

HANDBOOK OF THE INDIANS OF CALIFORNIA, A. L. Kroeber. Foremost American anthropologist offers complete ethnographic study of each group. Monumental classic. 459 illustrations, maps. 995pp. 5⅜ x 8½.
23368-5 Pa. $13.00

SHAKTI AND SHAKTA, Arthur Avalon. First book to give clear, cohesive analysis of Shakta doctrine, Shakta ritual and Kundalini Shakti (yoga). Important work by one of world's foremost students of Shaktic and Tantric thought. 732pp. 5⅜ x 8½. (Available in U.S. only)
23645-5 Pa. $7.95

AN INTRODUCTION TO THE STUDY OF THE MAYA HIEROGLYPHS, Syvanus Griswold Morley. Classic study by one of the truly great figures in hieroglyph research. Still the best introduction for the student for reading Maya hieroglyphs. New introduction by J. Eric S. Thompson. 117 illustrations. 284pp. 5⅜ x 8½.
23108-9 Pa. $4.00

A STUDY OF MAYA ART, Herbert J. Spinden. Landmark classic interprets Maya symbolism, estimates styles, covers ceramics, architecture, murals, stone carvings as artforms. Still a basic book in area. New introduction by J. Eric Thompson. Over 750 illustrations. 341pp. 8⅜ x 11¼.
21235-1 Pa. $6.95

GEOMETRY, RELATIVITY AND THE FOURTH DIMENSION, Rudolf Rucker. Exposition of fourth dimension, means of visualization, concepts of relativity as Flatland characters continue adventures. Popular, easily followed yet accurate, profound. 141 illustrations. 133pp. 5⅜ x 8½.
23400-2 Pa. $2.75

THE ORIGIN OF LIFE, A. I. Oparin. Modern classic in biochemistry, the first rigorous examination of possible evolution of life from nitrocarbon compounds. Non-technical, easily followed. Total of 295pp. 5⅜ x 8½.
60213-3 Pa. $4.00

PLANETS, STARS AND GALAXIES, A. E. Fanning. Comprehensive introductory survey: the sun, solar system, stars, galaxies, universe, cosmology; quasars, radio stars, etc. 24pp. of photographs. 189pp. 5⅜ x 8½. (Available in U.S. only)
21680-2 Pa. $3.75

THE THIRTEEN BOOKS OF EUCLID'S ELEMENTS, translated with introduction and commentary by Sir Thomas L. Heath. Definitive edition. Textual and linguistic notes, mathematical analysis, 2500 years of critical commentary. Do not confuse with abridged school editions. Total of 1414pp. 5⅜ x 8½. 60088-2, 60089-0, 60090-4 Pa., Three-vol. set $18.50

Prices subject to change without notice.

Available at your book dealer or write for free catalogue to Dept. GI, Dover Publications, Inc., 180 Varick St., N.Y., N.Y. 10014. Dover publishes more than 175 books each year on science, elementary and advanced mathematics, biology, music, art, literary history, social sciences and other areas.